Kyley Ewing
Editor

SPACETIME CONFERENCE 2024

Selected peer-reviewed papers presented at the Seventh International Conference on the Nature and Ontology of Spacetime, 16 - 19 September 2024, Albena, Bulgaria

MINKOWSKI
Institute Press

Kyley Ewing
Department of Philosophy
Cape Breton University
Canada

Front Cover: Aerial view of the famous Black Sea resort Albena (near Varna), Bulgaria (https://bnr.bg/en/post/101158323/albena-resort-celebrates-its-50th-anniversary)

ISBN: 978-1-998902-44-6 (softcover)
ISBN: 978-1-998902-45-3 (ebook)

Minkowski Institute Press
Montreal, Quebec, Canada
https://minkowskiinstitute.com/mip/

For information on all Minkowski Institute Press publications visit our website at
https://minkowskiinstitute.com/mip/books/

Most participants of the Seventh Spacetime Conference who stayed until the last day of the conference. Picture taken by Vladimir Petev (Front office manager of Hotel Sandy Beach).

CONTENTS

IV PASSAGE AND DIRECTION OF TIME   179

9 The happening of events in temporal relation to one another establishes bistable and selfsustaining dynamics of the Universe
Bruce M. Boman   181

10 Dynamical Structured Spacetime Realism (DSSR): The Emergence of Low-Dimensional CPT-Symmetric Spacetime
Charlie Dawson   211

PREFACE

This volume presents a selection of peer-reviewed papers from the Seventh International Conference on the Nature and Ontology of Spacetime that was held from September 16-19 in Albena, Bulgaria. The conference offered the opportunity for colleagues from around the world to meet and discuss some of the most pressing current questions facing both physicists and philosophers when it comes to the fundamental nature and ontology of spacetime. Held at the Sandy Beach Hotel on the breathtaking Black Sea, the conference provided a chance to not only discuss scientific and philosophical questions, but the opportunity to take in the beauty offered by Bulgarian nature.

Divided into talks and panel discussions, one of the highlights of the conference was the group conversations where participants came together to openly discuss, debate, and analyze topics such as 'becoming and the nature of time', 'the nature of gravitation', and 'open questions in spacetime physics'. The inquiring character of the discussions combined with the good-natured comradery between participants from both the scientific and philosophical disciplines showcased many of the best aspects of what it means to be part of a global community of theorists and thinkers.

Of particular note, the Seventh International Conference on the Nature and Ontology of Spacetime marked the 115th anniversary of the 1909 publication of Minkowski's 1908 lecture "Space and Time". Minkowski's lecture changed the course of the study of space and time insofar as it presented novel and groundbreaking ideas connected to the nature and ontology of the four-dimensional physics of the world. As a way to commemorate Minkowski's enduring impact and legacy, the conference included a special session that focused on Minkowski's actual and intended contributions to spacetime physics. Along similar lines, another main theme of the conference surrounded the question of what spacetime represents: Does spacetime represent the real, physical four-dimensional manifold that Minkowski advocated for or is spacetime merely a mathematical manifold?

Drawing exact boundaries between the topics discussed in the papers that make up the present volume is challenging. Nevertheless, for ease of presentation, the volume has been divided into four main sections. The first section 'Minkowski Spacetime' contains a selection of papers that celebrate Minkowski's contributions to spacetime physics and apply his ideas to current open questions in the physics and philosophy of spacetime. The second section 'Relativity and Cosmology' includes papers that aim to think of relativity from a new vantage point or reflect on the relationship between relativity and contemporary cosmology. The third section 'Ontology of Time' presents papers that question the picture of the world as a four-dimensional physical spacetime structure. The fourth section 'Passage and Direction of Time' addresses perennial and unresolved questions related to the nature of the passage and direction of time. It is hoped that readers of this volume come away with both a better understanding of and a few more questions about some of the most important and interesting ideas concerning the nature of our world.

Kyley Ewing
Assistant Professor, Philosophy
Cape Breton University

Contributors

Marcus Arvan
The University of Tampa
marvan@ut.edu

Prakash Bhat

Bruce M. Boman
Department of Mathematical Sciences
University of Delaware

Charlie Dawson
awesomechuckdawson@gmail.com

Călin Galeriu
Military Technical Academy "Ferdinand I"
Bucharest, Romania
calin.galeriu@mta.ro

Vesselin Petkov
Minkowski Institute
Montreal, Canada
vpetkov@minkowskiinstitute.com

Nikola Pirovski
Anthropology Laboratory
Department of Anatomy
Faculty of Medicine, Thracian University
Section Philosophy of Science and Problem Group on
Philosophy and Sociology of Medicine IFS-BAS
nikola.pirovski@trakia-uni.bg

Daniel Shanahan
Mariners Reach
Newstead, Queensland 4006

Anguel S. Stefanov
Corresponding member of the Bulgarian Academy of Sciences

Asher Yahalom
Department of Electrical & Electronic Engineering,
Faculty of Engineering, Ariel University, Israel
Center for Astrophysics, Geophysics, and
Space Sciences (AGASS), Ariel University, Israel
FEL User Center, Ariel University, Israel
asya@ariel.ac.il

Part I

MINKOWSKI SPACETIME

Kyley Ewing (Ed.), *Spacetime Conference 2024. Selected peer-reviewed papers presented at the Seventh International Conference on the Nature and Ontology of Spacetime, 16 - 19 September 2024, Albena, Bulgaria* (Minkowski Institute Press, Montreal 2025). ISBN 978-1-998902-44-6 (softcover), ISBN 978-1-998902-45-3 (ebook).

1 MINKOWSKI'S ACTUAL AND INTENDED CONTRIBUTIONS TO SPACETIME PHYSICS

VESSELIN PETKOV

Abstract   Although Hermann Minkowski, *de facto*, discovered and founded spacetime physics, his enormous contributions to fundamental physics have not been fully appreciated. The purpose of this chapter is two-fold – to state and explain his actual contributions to spacetime physics and to outline what his intended contributions might have been, if he had lived longer.

1   Introduction

In 2024, when we marked the 160th anniversary of the birth of Hermann Minkowski (1864-2024), it was natural to recall and reexamine his epoch-making contributions to spacetime physics. Given the depth of Minkowski's thinking and the far-reaching consequences of his results, it is also natural to wonder how fundamental (at least spacetime) physics might look like now if he had lived longer. In this sense, it is worth trying to imagine how Minkowski might have influence the advancement of spacetime physics.

As the title suggests the chapter is divided into two parts. First, I will discuss Minkowski's major and actual (not the widely shared misinterpreted) contribution to spacetime physics – the discovery of the spacetime structure of the world, i.e., the discovery of spacetime physics itself. It should be stressed that what Minkowski started to develop is a physics of a four-dimensional world, which is fundamentally different from the physics of the perceived by us three-dimensional world; it is precisely this research direction, outlined by Minkowski, that has not been appreciated (and probably never understood), let alone developed. I will also summarize the evidence that Minkowski arrived independently of Einstein at the equivalence of the times of observers in relative motion and independently of Poincaré at the conclusion that the Lorentz transformations can be regarded as rotations in a four-dimensional space.

Second, I will describe what Minkowski's intended contributions to spacetime physics might have been. What enables me to talk (not just speculate) about what Minkowski might have achieved is his explicitly outlined program of geometrizing physics – regarding four-dimensional physics, i.e. the physics of a four-dimensional world, merely as geometry of this four-dimensional world (spacetime), where material objects are a forever-given (static)

Kyley Ewing (Ed.), *Spacetime Conference 2024. Selected peer-reviewed papers presented at the Seventh International Conference on the Nature and Ontology of Spacetime, 16 - 19 September 2024, Albena, Bulgaria* (Minkowski Institute Press, Montreal 2025). ISBN 978-1-998902-44-6 (softcover), ISBN 978-1-998902-45-3 (ebook).

web of worldlines.

2   Minkowski's actual contributions to spacetime physics

115 years after Minkowski arrived at the stunning conclusion that the world is four-dimensional and developed the mathematical formalism of the four-dimensional physics of this world (what he called die Welt), his revolutionary achievements have not been properly acknowledged and appreciated.

One can hear physics students (and, disturbingly, even physicists[1]) talking about Einstein's spacetime of special relativity.

What is perhaps even worse is that the very mathematical formalism of spacetime, developed by Minkowski, is not clearly associated with his name. A rare exception is A. Ashtekar. In his Foreword to the second edition of the collection of papers *The Origin of Spacetime Physics* [2] he nicely describes the application of the developed by Minkowski four-dimensional mathematical formalism of spacetime physics to electrodynamics:

> The new edition includes a Chapter based on Minkowski's lecture to the Göttingen Scientific Society on December 21st, 1907, entitled *Fundamental Equations for the Electromagnetic Processes in Moving Bodies*. This is a much more detailed account of Minkowski's astonishingly deep understanding of how the fusion of space and time into a four-dimensional spacetime continuum leads to a reformulation of electrodynamics. In particular, this paper provides the tensorial formulation of Maxwell's equations and the action of the Lorentz group on the Maxwell field tensor and the source current. Because of its emphasis on four-dimensional geometry, this discussion of Maxwell's equations goes distinctly beyond Einstein's paper on *On the Electrodynamics of Moving Bodies*. Indeed, Minkowski's four-dimensional equations are exactly in the same form that we use today, more than a century later!

Before discussing Minkowski's actual contributions to spacetime physics let me first list them:

- Minkowski decoded the profound physical message hidden in the failed experiments (captured in Galileo's principle of relativity and the Michelson-Morley experiment) – all those experiments failed because the world is four-dimensional, which, when described in terms of the ordinary three-dimensional language (which, after Minkowski, turned our to be rather unscientific), means that inertial observers in relative motion have their own spaces and times; they perform experiments in their spaces and times

---

[1]Those who make such uninformed and unfair statements definitely have not read either Minkowski's or Einstein's papers. Such statements are particularly unfortunate because after Minkowski's 1908 world-view-changing lecture "Space and Time" Einstein had apparently had difficulty realizing the depth of Minkowski's ideas and his reaction to the developed by Minkowski four-dimensional physics had been rather hostile. Sommerfeld's recollection of what Einstein said on one occasion provides an indication of Einstein's initial attitude towards the work of his mathematics professor on the foundations of spacetime physics [1]: "Since the mathematicians have invaded the relativity theory, I do not understand it myself any more."

and therefore always find themselves to be at rest in their spaces (which explained the failure of all experiments to detect absolute motion).

- Minkowski showed that, in addition to the straightforward explanation of the negative results of the experiments performed to discover absolute motion, the four-dimensional world also provides not only a natural, but also the only possible explanation[2] of length contraction.

- Minkowski, *de facto*, showed, but did not state it explicitly, that neither those negative-result-experiments nor length contraction would be possible if the world were *not* four-dimensional, because those results and length contraction turned out to be manifestations of the four-dimensionality of the world; length contraction is a manifestation of the reality of the worldtube of the contracting body, which, in turn, demonstrates that reality is a four-dimensional world (which we call spacetime).

- Minkowski realized that the physics of a four-dimensional world is profoundly different from the physics of a three-dimensional world – in a three-dimensional world there are particles moving with constant velocity or with acceleration, whereas in a four-dimensional world (spacetime) there are straight and curved worldlines. That is why:

  - in a three-dimensional world physics deals with motion and interactions of three-dimensional bodies.
  - in a four-dimensional world physics deals with interrelations between worldlines (or worldtubes in the case of extended bodies) and deformations[3] of curved worldtubes because there is no dynamics in spacetime; it is incorrect even to say that a particle follows a timelike worldline because in spacetime the particle itself is a timelike worldline.

Let me now explain how Minkowski decoded the profound physical message hidden in the unsuccessful experimental attempts to detect absolute motion (captured in Galileo's principle of relativity and the Michelson-Morley experiment), which led him to the idea of the spacetime structure of the world. He revealed that hidden message by analyzing both those experiments and the internal logic of the mathematical formalism of classical mechanics.

First, Minkowski arrived independently from Einstein[4] at the *equivalence* of the time t of a stationary observer and a *second* time t', which Lorentz introduced (as "an auxiliary mathematical quantity"[5] to explain the Michelson-Morley experiment) calling it the *local*

---

[2]Why this is so is explained in [3, pp. 12-13].

[3]As we will see, Minkowski showed that the acceleration of a body is, in spacetime, the curvature (rather the *deformation*) of the worldtube of the accelerating body.

[4]See the last part of this section.

[5]Lorentz admitted his failure to recognize the profound physical meaning of his idea to introduce a second time in physics (and that a second time is impossible in a three-dimensional world as Minkowski showed). In 1916 in a note added to the second edition of his *The Theory of Electrons and Its Applications to the Phenomena of Light and Radiant Heat* he wrote [4]:

The chief cause of my failure was my clinging to the idea that the variable t only can be *time* of a moving observer (whose x' axis is along the x axis of the stationary observer). He figured out, on the basis of the Lorentz transformations, that t and t' should be treated equally, which implies that *there exist more then one time in the physical world*. Then, as a mathematician, Minkowski immediately realized that not only do observers in relative motion have different times, but they also have different spaces [5, p. 62]:

> One can call t' time, but then must necessarily, in connection with this, define space by the manifold of three parameters x', y, z in which the laws of physics would then have exactly the same expressions by means of x', y, z, t' as by means of x, y, z, t. Hereafter we would then have in the world no more *the* space, but an infinite number of spaces analogously as there is an infinite number of planes in three-dimensional space. Three-dimensional geometry becomes a chapter in four-dimensional physics. You see why I said at the beginning that space and time will recede completely to become mere shadows and only a world in itself will exist.[6]

Minkowski specifically pointed out that [5, p. 65]

> Neither Einstein nor Lorentz disputed the concept of space... To go beyond the concept of space in such a way is an instance of what can only be imputed to the audacity of mathematical culture.

So Minkowski's mathematical analysis led him to a four-dimensional mathematical structure (his *die Welt*). When he assumed that this structure represented a real four-dimensional world it became immediately evident why it is impossible to detect experimentally absolute motion – in this four-dimensional world there is no absolute space (such a space can exist only in a three-dimensional world) with respect to which bodies move; observers in relative motion have different spaces and times (when the four-dimensional world is described in terms of the ordinary three-dimensional language) and every time they perform experiments to detect their motion with respect to the assumed absolute space they do it in their own spaces by using their own times and always determine that they are at rest with respect to their own spaces. Then Minkowski realized that the physical message, which all failed experiments have been trying to convey to us for centuries, is indeed of profound depth – those experiments all failed because absolute space does not exist since the physical world is four-dimensional.

The way Minkowski realized that the physical world must be four-dimensional in order that absolute motion does not exist, which provided a straightforward explanation of why all experiments to detect it failed, makes it clear why he insisted

---

considered as the true time, and that my local time t' must be regarded as no more than an auxiliary mathematical quantity.

[6] Here Minkowski mentions the beginning of his lecture [5, p. 57]:

The views of space and time which I want to present to you arose from the domain of experimental physics, and therein lies their strength. Their tendency is radical. From now onwards space by itself and time by itself shall completely fade into mere shadows and only a specific union of the two will still stand independently on its own.

at the beginning of his 1908 lecture that the strength of the new views of space and time comes from the fact that they "arose from the domain of experimental physics."

Due to its importance for the foundations of spacetime physics, let me stress it: Minkowski's discovery of the spacetime structure of the world *explained* the physical meaning of the principle of relativity (the impossibility to discover absolute uniform motion and absolute rest), which Einstein merely *postulated* – all physical phenomena look in the same way to observers in uniform relative motion (so they cannot tell who is moving as the experimental evidence proved) *because* they have different times and different spaces. Each observer performs experiments in his own space and time and for this reason the physical phenomena look in the same way to all observers, e.g., the speed of light is the same for them since each observer measures it in his own space by using his own time.

What has been virtually ignored (not just unappreciated) for over a century is Minkowski's major contribution to fundamental physics – the discovery of the spacetime structure of the world, i.e. the discovery that the physical realty is a four-dimensional world (which we call spacetime). The four-dimensional formalism developed by Minkowski for the description of this four-dimensional world (spacetime) has been adopted in physics, but most (probably all) physicists do not ask themselves the fundamental question, which the mathematician Minkowski asked – What does this four-dimensional mathematical formalism represent? What is worse is that many physicists regard the introduced by Minkowski four-dimensional world (spacetime) as nothing more than "an abstract four-dimensional mathematical continuum" [6].

To see that such a view directly contradicts the existing relativistic experimental evidence, let me explicitly formulate Minkowski's argument for the reality of spacetime – the experiments, which failed to discover absolute motion (from Galileo's to Minkowski's time, including the Michelson-Morley experiment), *would be impossible* (i.e., they would discover absolute motion) *if spacetime were not real*. To see why, assume that spacetime is indeed an abstract space and that reality is what we perceive – an evolving in time three-dimensional world. Then there would exist a *single* space (since a three-dimensional world allows the existence of *one* space), which as such would be absolute (the same for all observers). As a space constitutes a class of simultaneous events (the space points at a given moment), a single (absolute) space implies absolute simultaneity and therefore absolute time as well. Hence, a three-dimensional world allows *only* absolute space and absolute time in contradiction with the experimental evidence that uniform motion with respect to the absolute space cannot be discovered as encapsulated in the principle of relativity: as Minkowski explained it, the reason for the failure of all experiments to detect absolute motion was the non-existence of an absolute (i.e., a single) space; if such a space existed (i.e., if the world were three-dimensional), uniform motion relative to that space would be detected. I hope it is now clear why Minkowski's main argument, taken even *alone*, effectively proves the reality of spacetime.

Minkowski did not state explicitly that absolute motion would be discovered if the world were *not* four-dimensional, most probably because it looked completely obvious to him and therefore unnecessary to be included in his dense 1908 lecture.

As Minkowski's main argument demonstrates that many spaces (i.e., relative simultaneity) are possible only in a four-dimensional world,[7] it also effectively proves that length contraction and time dilation are possible only in a four-dimensional world because these two relativistic effects are specific manifestations of the fact that observers in relative motion have *different* spaces (different sets of simultaneous events, i.e., relative simultaneity). To see that this is really the case, let us examine Minkowski's explanation of length contraction as presented graphically in Fig. 1.

Minkowski considered two bodies in uniform relative motion represented by their red and green worldtubes[8] as shown on the right half of Fig. 1. He demonstrated that, like the failed experiments to detect absolute motion are manifestations of the reality of spacetime, length contraction is another such manifestation.

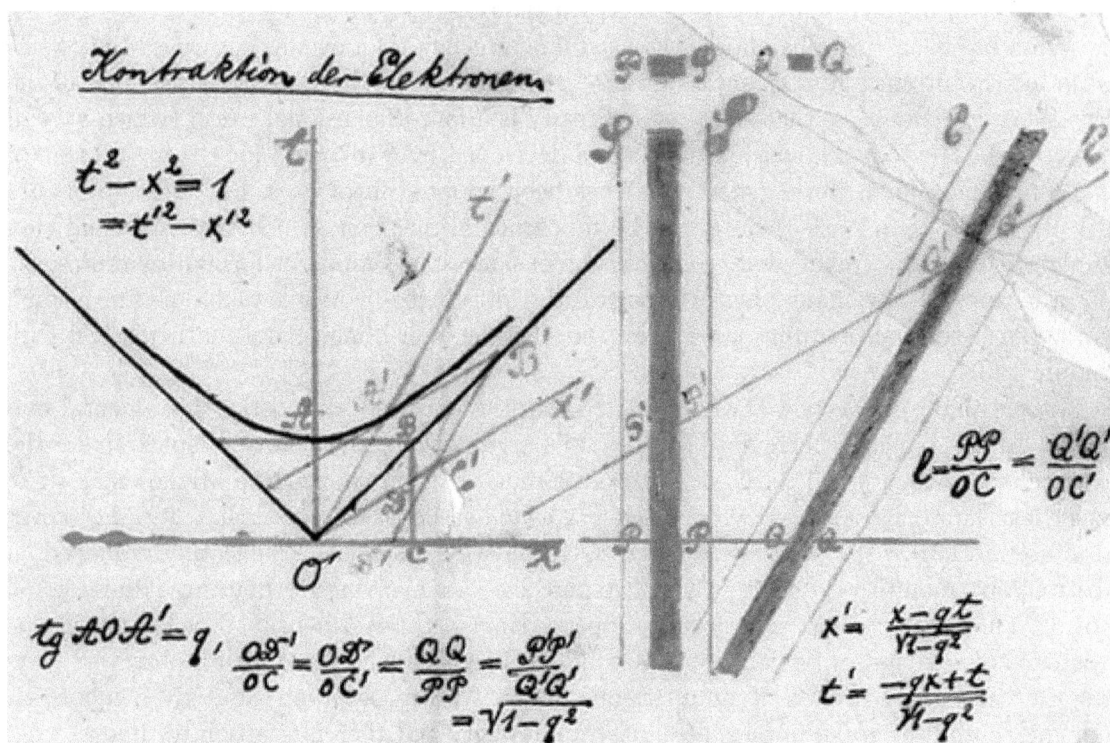

Figure 1: The transparency which Minkowski used at his lecture in Cologne on September 21, 1908. It shows Fig. 1 in his paper " Space and Time." Source (with permission): Cover of *The Mathematical Intelligencer*, Volume 31, Number 2 (2009).

Consider the body represented by the vertical red worldtube. The three-dimensional cross-section PP, resulting from the intersection of the body's worldtube and the space (represented by the red horizontal line in Fig. 1) of an observer at rest with respect to the body, is the body's proper length. The three-dimensional cross-section $P'P'$, resulting from the intersection of the red body's worldtube and the space (represented by the inclined green line) of an observer at rest with respect to the green body (represented by the inclined green worldtube), is the relativistically contracted length of the red body measured by

---

[7]Rigorously speaking, many spaces are possible at least in a four-dimensional world.

[8]Minkowski used the term strip, not worldtube.

that observer (the cross-section $P'P'$ only appears longer than PP because a fact of the pseudo-Euclidean geometry of spacetime is represented on the Euclidean surface of the page). Note that while measuring the *same* body, the two observers measure *two different* three-dimensional bodies represented by the cross-sections PP and $P'P'$ in Fig. 1 (this relativistic situation will not be truly paradoxical only if what is meant by "the same body" is the body's worldtube).

To see that length contraction is impossible in a three-dimensional world,[9] assume, just for the sake of the argument, that the worldtube of the red body did not exist as a four-dimensional object and were nothing more than an abstract geometrical construction. Then, what would exist would be a single three-dimensional body, represented by the cross-section PP, the red and green body would share the *same* three-dimensional space (in Fig. 1 the red and green line would coincide) and both observers, associated with the red and green bodies, respectively, would measure the *same* three-dimensional body of the *same* length because nothing else would exist. Therefore, not only would length contraction be *impossible*, but relativity of simultaneity would be also impossible since a spatially extended three-dimensional object is defined in terms of *simultaneity* – all parts of a body taken *simultaneously* at a given moment – and as both observers in relative motion would measure the same three-dimensional body (represented by the cross-section PP) they would share the *same* class of simultaneous events in contradiction with relativity.

Let us now summarize Minkowski's specific argument for the reality of spacetime: the relativistic length contraction of a body is a manifestation of the reality of the body's world-tube and therefore another manifestation of the reality of spacetime – (i) the red and the green bodies must have *different* three-dimensional spaces (i.e., relative simultaneity, which is impossible in a three-dimensional world) in order to intersect the red body's worldtube at two *different* places (in the cross-sections PP and $P'P'$), and (ii) the red body's worldtube must be real in order to be intersected at two *different* places.

Another relativistic effect – time dilation – is also such a manifestation of relativity of simultaneity (the fact that observers in relative motion have different three-dimensional spaces); that is, a manifestation of the reality of spacetime. In other words, like length contraction, time dilation is also impossible in a three-dimensional world: in order that time dilation be possible (i) two observers in relative motion, measuring this relativistic effect, must have different three-dimensional spaces, and (ii) the worldtubes of the clocks involved in the measurement must be real four-dimensional objects (for details see [7]).

As length contraction and especially time dilation have been repeatedly confirmed by experiment[10] *none of these experiments would be possible if spacetime were not real.*

---

[9]Minkowski did not state this in his 1908 lecture most probably for the same reason (as in the case of his main argument) – it is exceeding obvious especially to a mathematician.

[10]Length contraction was tested experimentally, along with time dilation, by the muon experiment in the muon reference frame [8]:

> In the muon's reference frame, we reconcile the theoretical and experimental results by use of the length contraction effect, and the experiment serves as a verification of this effect.

In order to appreciate fully Minkowski's contributions to spacetime physics it should be emphasized that at least two things appear to indicate that he arrived independently at what Einstein called special relativity and at the concept of spacetime, but Einstein and Poincaré published first while Minkowski had been developing the four-dimensional formalism of spacetime physics reported on 21 December 1907 and published in 1908 as a 59-page treatise "The Fundamental Equations for Electromagnetic Processes in Moving Bodies" (translated in English in [2]).

The first indication is the novelty and depth of Minkowski's approach and theoretical achievements (contained in his two papers included in [2]), which demonstrate that he was developing his own original insights, not interpreting someone's results. Indeed, in his paper "Space and Time" Minkowski presented his revolutionary ideas of regarding physics as *geometry* of the discovered by him four-dimensional world and in the treatise "The Fundamental Equations for Electromagnetic Processes in Moving Bodies" single-handedly developed the four-dimensional mathematical formalism of spacetime physics (neither Einstein not Poincaré in their 1905 papers were even close to it).

Also, Minkowski's way of doing physics is profoundly different from Einstein's and Poincaré's ways, which additionally demonstrates Minkowski's independent path to the discovery of the spacetime structure of the world. For example Einstein used postulates[11] – the relativity postulate and the postulate of the constancy of the speed of light – without even attempting to *explain* them, whereas Minkowski provided *explanations* (for details, see the first two chapters of [3]). Einstein believed that the essence of his special relativity (and later of his general relativity) was the *relativity* of physical quantities, whereas Minkowski, as a mathematician, searched for and found the underlying *absolute* entity (spacetime) that makes possible the very existence of relative quantities[12] (as mentioned above relativity of space and time is impossible in a three-dimensional world). Einstein talked about relativity of simultaneity, whereas Minkowski showed that if inertial observers in relative motion have different times (as Einstein postulated), they have different (three-dimensional) spaces as well.[13] Another indication that Minkowski arrived

---

[11]Minkowski also suggested to use a postulate – the postulate of the absolute world [5]):

> I think the word *relativity postulate* used for the requirement of invariance under the group G_c is very feeble. Since the meaning of the postulate is that through the phenomena only the four-dimensional world in space and time is given, but the projection in space and in time can still be made with certain freedom, I want to give this affirmation rather the name *the postulate of the absolute world* (or shortly the world postulate).

However, the world postulate is a fundamentally different kind of postulate – it simply states the *dimensionality* of the world as revealed by Minkowski's rigorous analysis of the experiments captured in the relativity postulate, whereas the relativity postulate states that uniform motion with respect to space cannot be detected, which needs to be explained *why*; the question of *why* the world is four-dimensional is a much deeper question, perhaps as fundamental as the question "Why does the world exist?"

[12]Mathematicians are well-aware that relative quantities are descriptions of something absolute: "The emphasis on the *geometry* means an emphasis on the *absolutes* which underlie relative descriptions" [9].

[13]As a mathematician Minkowski probably realized that immediately and remarked (as indicated above) "Neither Einstein nor Lorentz disputed the concept of space"), i.e., neither of them stated that inertial observers in relative motion should also have different spaces. Einstein discussed relativity of simultaneity, but seems to have not realized in his early papers that a class of simultaneous events forms a three-dimensional space which appears to explain why Einstein mentioned that inertial observers in relative motion have

independently at the conclusion that the times of inertial observers in relative motion have the same status is that he never mentioned relativity of simultaneity, although he had been undoubtedly aware that space is defined in terms of simultaneity – the class of all space points at a given moment of time (i.e. simultaneous with that moment) – and therefore different spaces imply different classes of simultaneous events.

The second indication are Max Born's recollections about Minkowski's work shared during a seminar in 1905 and about his conversations with Minkowski.

By 1905 Hermann Minkowski was already internationally recognized as an exceptional mathematical talent. At that time he became interested in the electron theory and especially in an unresolved issue at the very core of fundamental physics – at the turn of the nineteenth and twentieth century Maxwell's electrodynamics had been interpreted to show that light is an electromagnetic wave, which propagates in a light-carrying medium (the luminiferous ether), but its existence was put into question since the Michelson-Morley interference experiments failed to detect the Earth's motion in that medium.

Minkowski's documented involvement with the electrodynamics of moving bodies began in the early summer of 1905 when he and his friend David Hilbert co-directed a seminar in Göttingen on the electron theory. The paper of Minkowski's student Albert Einstein "On the Electrodynamics of Moving Bodies" (this volume) was not published at that time; *Annalen der Physik* received Einstein's paper on June 30, 1905. Poincaré's longer paper "On the Dynamics of the Electron" (this volume), in which Poincaré regarded the Lorentz transformations as rotations in a four-dimensional space with time as the fourth dimension, was not published either; *Rendiconti del Circolo matematico di Palermo* received Poincaré's paper on July 23, 1905. Also, "Lorentz's 1904 paper (with a form of the transformations now bearing his name) was not on the syllabus" [11].

Minkowski's student Max Born, who attended the seminar in 1905, wrote [12]:

> We studied papers by Hertz, Fitzgerald, Larmor, Lorentz, Poincaré, and others but also got an inkling of Minkowski's own ideas which were published only two years later.

Born also recalled what Minkowski had specifically mentioned a number of times during the seminar in 1905 [13]:

> I remember that Minkowski occasionally alluded to the fact that he was engaged with the Lorentz transformations, and that he was on the track of new interrelationships.

Again Born wrote in his autobiography about what he had heard from Minkowski after Minkowski's lecture "Space and Time" given on September 21, 1908 [14]:

> He told me later that it came to him as a great shock when Einstein published his paper in which the equivalence of the different local times of observers moving

---

different spaces hardly in the fifth appendix to his popular book *Relativity: The Special and General Theory* which was added in 1952 [10, p. 103]: "But it must now be remembered that there is an infinite number of spaces, which are in motion with respect to each other." The correct expression is: those spaces only *appear* to be in relative motion, because they are merely different *imaginary* three-dimensional cross-sections of spacetime, which is not objectively divided into spaces.

relative to each other were pronounced; for he had reached the same conclusions independently but did not publish them because he wished first to work out the mathematical structure in all its splendour. He never made a priority claim and always gave Einstein his full share in the great discovery.

Minkowski asked Einstein to send him the 1905 paper "On the Electrodynamics of Moving Bodies" hardly on October 9, 1907 [15].

3   Minkowski intended contributions

By intended contributions I mean what Minkowski might have achieved, had he lived longer, by employing his program of geometrizing physics, i.e., by regarding physics as geometry of spacetime. Let me again first list what I regard as most probable results Minkowski's program might have produced:

- Minkowski might have realized that inertia is another manifestation of the reality of spacetime, i.e., of the four-dimensionality of the world. Given that he regarded reality as a four-dimensional world (his die *Welt*), which means that physical bodies are four-dimensional worldtubes, it appears virtually self-evident that the curved (*deformed*) worldtube of an accelerating body *resists* its deformation (like a deformed three-dimensional rod resists its deformation) – it is this static four-dimensional resistance, which we observe (and measure) as the resistance a body offers to its acceleration and which we call the body's inertia.

- The employment of Minkowski's program of geometrizing physics (and the explicit assumption that the worldtubes of bodies are real four-dimensional objects) to an analysis of the experimental fact (known since Galileo) that bodies of different masses fall toward the Earth's surface with the same acceleration, might have led him to the realization that gravitational phenomena are manifestations of a *real non-Euclidean* four-dimensional world.

After Minkowski explained in his lecture *Space and Time* that the true reality is a four-dimensional world in which all ordinarily perceived three-dimensional particles are a forever given web of worldlines, he outlined his ground-breaking idea of regarding physics as spacetime geometry [5, p. 112]:

> The whole world presents itself as resolved into such worldlines, and I want to say in advance, that in my understanding the laws of physics can find their most complete expression as interrelations between these worldlines.

Then he started to implement his program by explaining that inertial motion is represented by a *straight* timelike worldline, after which he pointed out that [5, p. 115]:

> With appropriate setting of space and time the substance existing at any world-point can always be regarded as being at rest.

In this way he explained not only *why* the times of inertial observers are equivalent (their times can be chosen along their timelike worldlines and all straight timelike worldlines in spacetime are equivalent) but also the physical meaning of the relativity principle (as discussed in the previous section) – the physical laws are the same for all inertial observers (inertial reference frames), i.e. all physical phenomena look exactly the same for all inertial observers, because every observer describes them in his own space (in which he is at rest) and uses his own time. For example the speed of light is the same for all observers because each observer measures it in its own space using his own time.

Then Minkowski explained that accelerated motion is represented by a *curved* or, more precisely, *deformed* worldline and noticed that "Especially the concept of *acceleration* acquires a sharply prominent character" [5, p. 66].

As Minkowski knew that a particle moving by inertia offers no resistance to its motion with constant velocity (which explains why inertial motion cannot be detected experimentally as Galileo first demonstrated), whereas the accelerated motion of a particle can be discovered experimentally since the particle *resists* its acceleration, he might have very probably linked the sharp physical distinction between inertial (non-resistant) and accelerated (resistant) motion with the sharp geometrical distinction between inertial and accelerated motion represented by straight and deformed (curved) worldlines, respectively.

The realization that an accelerated particle (which resists its acceleration) is a deformed worldtube in spacetime would have allowed Minkowski to notice two virtually obvious implications of this spacetime fact:

- The acceleration of a particle is absolute not because it accelerates with respect to some absolute space, but because its worldtube is deformed, which is an absolute geometrical and physical fact.

- The resistance a particle offers to its acceleration (i.e. its inertia) appears to be linked to its *deformed* worldtube. That is, the inertial force with which the particle resists its acceleration seems to be arising from a static restoring force in the *deformed* worldtube of the accelerated particle.

As we see Minkowski had actually made a gigantic step toward revealing the origin and nature of inertia by showing that a curved (deformed) timelike worldline in spacetime represents an accelerated particle and explained why "the concept of *acceleration* acquires a sharply prominent character" – because the acceleration of a particle is proportional to the curvature of its worldline [5, p. 69].

Linking the acceleration of a particle with the curvature (*deformation*) of its worldline (rather worldtube) leads to the almost self-evident conjecture mentioned above [7, Ch. 9] – the *deformation* of the particle's worldtube does appear to give rise to a static restoring force (statically resisting the worldtube's deformation), which we interpret as the inertial force (with which the particle resists its acceleration).

Three facts strongly suggest the conjecture that *the worldtube of an absolutely accelerated body statically resists its deformation*[14] *and this resistance manifests itself as the body's inertia*:

---

[14]It should be stressed that it is the *deformation* of a worldline (not its curvature) that seems to be responsible for an inertial force, because it is the *deformation* of a worldline that manifests itself as an

- it is an experimental fact (explicitly discussed by Newton[15]) that a body resists its absolute acceleration by exerting an inertial force on the obstacle that prevents it from moving by inertia

- the worldtube of such a body is *deformed*

- since Minkowski regarded the worldtube of a three-dimensional body as *real*[16] it follows that it should resist its deformation (like a three-dimensional rod resists its deformation).

Now (in the 21st century) we can see that the static resistance "arising" in a deformed worldtube (and the involved four-dimensional stress) can be traced down to the self-forces acting on the accelerated constituents (elementary particles) of the accelerated body represented by that worldtube, which have contributions from electromagnetic, weak and strong interactions [7, Ch. 9].

Had Minkowski lived longer he might have been thrilled to realize that inertia is another manifestation of the reality of spacetime, i.e., of the four-dimensionality of the world (like length contraction, known at his time, and like time dilation and the twin paradox [7, Ch. 5]).

To demonstrates the enormous potential of Minkowski's program of geometrizing physics let us assume that, in addition to his knowledge of the experimental fact that accelerating bodies resist their acceleration, Minkowski had read Galileo's works, particularly Galileo's analysis demonstrating that heavy and light bodies fall at the *same* rate [16]. In this analysis Galileo virtually came to the conclusion (which is now a well-established experimental fact) that a falling body does not resist its fall [16]:

> But if you tie the hemp to the stone and allow them to fall freely from some height, do you believe that the hemp will press down upon the stone and thus accelerate its motion or do you think the motion will be retarded by a partial upward pressure? One always feels the pressure upon his shoulders when he prevents the motion of a load resting upon him; but if one descends just as rapidly as the load would fall how can it gravitate or press upon him? Do you not see that this would be the same as trying to strike a man with a lance when he is running away from you with a speed which is equal to, or even greater, than that with which you are following him? You must therefore conclude that,

---

absolute acceleration in both flat and curved spacetime; a curved worldline in flat spacetime is always deformed, but in curved spacetime a geodesic worldline is naturally curved due to the curvature of spacetime and is not deformed. The worldline of an absolutely accelerated particle in curved spacetime is not geodesic since it is additionally curved, i.e., it is deformed.

[15]"Inherent force of matter is the power of resisting by which every body, as far as it is able, perseveres in its state either of resting or of moving uniformly straight forward," Isaac Newton, *The Principia: Mathematical Principles of Natural Philosophy*. A New Translation by I. Bernard Cohen and Anne Whitman assisted by Julia Budenz (University of California Press, Berkeley 1999), p. 404.

[16]Those tempted to question this fact should return to Minkowski's argument (discussed above) for the reality of the worldtube of a relativistically contracted body to determine whether they are indeed questioning an experimental fact.

during free and natural fall, the small stone does not press upon the larger and consequently does not increase its weight as it does when at rest.

Then the path to the idea that gravitational phenomena are manifestations of the curvature of spacetime would have been open to Minkowski – the experimental fact that a falling particle accelerates (which means that its worldtube is curved), but offers no resistance to its acceleration (which means that its worldtube is not deformed) can be explained only if the worldtube of a falling particle is *both curved and not deformed*, which is impossible in the flat Minkowski spacetime where a curved worldtube is always deformed. Such a worldtube can exist only in a non-Euclidean spacetime whose geodesics are naturally curved due to the spacetime curvature, but are not deformed.

Since Minkowski regarded spacetime (*die Welt*) as real, it would not have been difficult for him (as a mathematician who listens to what the mathematical formalism tells him and is not affected by the appearance that gravitation looks like a physical interaction) to realize that gravitational phenomena are fully explained as manifestations of the non-Euclidean geometry of spacetime with no need to assume the existence of gravitational interaction. Indeed, particles fall toward the Earth's surface and planets orbit the Sun not due to a gravitational force or interaction, but because they move by inertia (non-resistantly); expressed in correct spacetime language, the falling particles and planets are geodesic worldlines (or rather worldtubes) in spacetime.

Minkowski would have easily explained the force acting on a particle on the Earth's surface, i.e. the particle's weight. The worldtube of a particle falling toward the ground is geodesic, which, in ordinary language, means that the particle moves by inertia (non-resistantly). When the particle lands on the ground it is prevented from moving by inertia and it resists the change of its inertial motion by exerting an inertial force on the ground. Like in flat spacetime the inertial force originates from the *deformed* worldtube of the particle which is at rest on the ground.[17] So the weight of the particle that has been traditionally called gravitational force would turn out to be inertial force, which naturally explains the observed equivalence of inertial and gravitational forces. While the particle is on the ground its worldtube is deformed (due to the curvature of spacetime), which means that the particle is being constantly subjected to a curved-spacetime acceleration (keep in mind that acceleration means deformed worldtube!); the particle resists its acceleration through the inertial force and the measure of the resistance the particle offers to its acceleration is its inertial mass, which traditionally has been called (passive) gravitational mass. This fact naturally explains the equivalence between a particle's inertial and gravitational masses, which turned out to be the same thing.

In this way, Minkowski would have again *explained* one more set of experimental facts (the equivalence of the effects of acceleration and gravitation) which Einstein merely *postulated* – Einstein "explained" these experimental facts by his equivalence postulate (i.e., the principle of equivalence). So Minkowski would have explained Einstein's equivalence postulate exactly like he explained Einstein's relativity postulate (i.e., the principle of relativity).

---

[17]Note again that Minkowski would have explained this fact only because he regarded spacetime as real – a fact deduced from all failed *experiments* designed to detect absolute uniform motion.

4 Conclusion

Despite the fact that Hermann Minkowski discovered and single-handedly[18] developed space-time physics, his enormous contributions have not been fully appreciated. That is why, Minkowski's actual contributions to spacetime physics were listed and discussed in the first part of the chapter. The second part explored the logical implications of Minkowski's program of geometrizing physics and, on the basis of that, outlined what his intended contributions might have been, if he had lived longer.

References

[1] A. Sommerfeld, "To Albert Einstein's Seventieth Birthday." In: *Albert Einstein: Philosopher-Scientist.* P. A. Schilpp, ed., 3rd ed. (Open Court, Illinois 1969) pp. 99-105, p. 102.

[2] V. Petkov (ed.), *The Origin of Spacetime Physics*, 2nd ed., with a Foreword by A. Ashtekar (Minkowski Institute Press, Montreal 2023).

[3] V. Petkov, *Seven Fundamental Concepts in Spacetime Physics*, 2nd ed. SpringerBriefs in Physics (Springer, Heidelberg 2024).

[4] H. A. Lorentz, *The Theory of Electrons and Its Applications to the Phenomena of Light and Radiant Heat*, 2nd ed. (Dover, Mineola, New York 2003), p. 57; see also his comment on p. 321.

[5] H. Minkowski, Space and Time, in [2, p. 148].

[6] N. D. Mermin, What's Bad About This Habit? *Physics Today* **62** (5) 2009, p. 8.

[7] V. Petkov, *Relativity and the Nature of Spacetime*, 2nd ed. (Springer, Heidelberg 2009), Ch. 5.

[8] G. F. R. Ellis and R. M. Williams, *Flat and Curved Space-Times* (Oxford University Press, Oxford 1988).

---

[18]This is completely evident from his 59-page treatise "The Fundamental Equations for Electromagnetic Processes in Moving Bodies" (published in 1907), where Minkowski, indeed, single-handedly developed the four-dimensional mathematical formalism of the physics of the discovered by him four-dimensional world (die *Welt*), which we now call spacetime physics. Neither Einstein nor Poincaré in their 1905 papers (and later) were even close to Minkowski's epoch-making discovery. In his paper "On the Dynamics of the Electron" (published in 1906 in *Rendiconti del Circolo matematico di Palermo*) Poincaré regarded the Lorentz transformations as rotations in a four-dimensional space with time as the fourth dimension, but did not even attempt to develop its mathematical formalism, maybe because he regarded that four-dimensional space as nothing more than a mathematical space, which does not represent a real four-dimensional world. As far as Einstein is concerned, here is again Sommerfeld's recollection [1] (quoted in the begining of Sec. 2) according to which Einstein's initial attitude towards the work of his mathematics professor on the foundations of spacetime physics had been rather hostile: "Since the mathematicians have invaded the relativity theory, I do not understand it myself any more."

[9] E.G. Peter Rowe, *Geometrical Physics in Minkowski Spacetime* (Springer, London 2001), p. vi.

[10] A. Einstein, *Relativity: The Special and General Theory*, new publication in the collection of five works by Einstein: A. Einstein, *Relativity*, ed. by V. Petkov (Minkowski Institute Press, Montreal 2018), p. 109.

[11] S. Walter, Minkowski, Mathematicians, and the Mathematical Theory of Relativity, in H. Goenner, J. Renn, J. Ritter, T. Sauer (eds.), *The Expanding Worlds of General Relativity, Einstein Studies*, volume 7, (Birkhäuser, Basel 1999) pp. 45-86, p. 46.

[12] M. Born, *Physics in My Generation* 2nd ed. (Springer-Verlag, New York 1969) p. 101.

[13] Quoted from T. Damour, "What is missing from Minkowski's 'Raum und Zeit' lecture," *Annalen der Physik* **17** No. 9-10 (2008), pp. 619-630, p. 626.

[14] M. Born, *My Life: Recollections of a Nobel Laureate* (Scribner, New York 1978) p. 131.

[15] Postcard: Minkowski to Einstein, October 9, 1907, in: M.J. Klein, A. J. Kox, and R. Schulmann (eds) *The Collected Papers of Albert Einstein*, Volume 5: *The Swiss Years: Correspondence*, 1902-1914 (Princeton University Press, Princeton 1995), p. 62.

[16] Galileo, *Dialogues Concerning Two Sciences*. In: S. Hawking (ed.), *On The Shoulders Of Giants*, (Running Press, Philadelphia 2002) pp. 399-626, p. 447.

2 THE ILLUSION OF ACCELERATION IN THE RETARDED LIÉNARD-WIECHERT ELECTROMAGNETIC FIELD

Călin Galeriu

Abstract   It is generally assumed that the retarded Liénard-Wiechert electromagnetic field produced by a point particle depends on the acceleration of that source particle. This dependence is not real, it is an illusion. The true electromagnetic interaction is time symmetric (half retarded and half advanced) and depends only on the positions and velocities of the electrically charged particles. A different acceleration of the retarded source particle will result in a different position and velocity of the advanced source particle, changing in this way the Lorentz force felt by the test particle.

1   Introduction

While the algebraic formalism of classical electrodynamics is well established, we would also like to have a comprehensive geometrical derivation of the electrodynamic forces between electrically charged point particles. From a geometrical point of view, the interacting particles are represented by worldlines in Minkowski space, and we hope that, with the proper theoretical structure in place, "the laws of physics can find their most complete expression as interrelations between these worldlines"[1].

We develop our theory by relying on geometrical intuition, as well as on physical and philosophical arguments. We start with the simplest case of electrostatic interaction, then we allow for constant velocities, then we allow for constant accelerations, and finally we allow for variable accelerations. Along the way we introduce new postulates, expanding the theoretical framework, until we end up with a time symmetric action-at-a-distance theory that reproduces the classical electrodynamics theory as a second order approximation.

2   Source particle at rest

Consider a source particle at rest, with electric charge Q, and a field point at $\vec{R}$, where $\vec{R}$ is the displacement vector from the source particle to the field point. The particle is the

Kyley Ewing (Ed.), *Spacetime Conference 2024. Selected peer-reviewed papers presented at the Seventh International Conference on the Nature and Ontology of Spacetime, 16 - 19 September 2024, Albena, Bulgaria* (Minkowski Institute Press, Montreal 2025). ISBN 978-1-998902-44-6 (softcover), ISBN 978-1-998902-45-3 (ebook).

source of an electric field $\vec{E}$ that points in the radial direction of $\vec{R}$ and has the magnitude of Q/R^2 (in Gaussian units), where R is the distance between the source particle and the field point. In this reference frame the magnetic field $\vec{B}$ is zero and the Lorentz force has only an electric part, that is $\vec{F} = q\vec{E}$.

A Lorentz transformation could bring us to a reference frame where the source particle is moving with constant velocity, the magnetic field is not zero, and the Lorentz force, for a test particle in motion, has also a magnetic part. This Lorentz transformation in effect introduces the velocity dependent magnetic interaction as a relativistic efect. However, even in this case, the electromagnetic interaction is due to the same four-force as before, which has different spatial and temporal components in different inertial reference frames [2]. For this reason in Section 2 we only investigate the case of a source particle at rest.

2.1   Test particle at rest

Consider a test particle at rest, with electric charge q.

When the interaction is instantaneous, as in classical Newtonian physics, the direction of the force $\vec{F}$ is along the straight line connecting the two particles. Let this also be the direction of the x axis of our reference frame. Without loss of generality we assume that both particles have a positive x coordinate, with the test particle being farther away from the origin. We also assume that $Q > 0$ and $q < 0$, such that the electrostatic force is attractive. This situation is represented graphically in Figure 1 (a), where the test particle is at A, the source particle is at M, and the segment AM has length R.

When the interaction is retarded, and the 3D Euclidean space is embedded into the 4D Minkowski space, the direction of the four-force $\mathbf{F} = (\vec{F}, 0)$ no longer matches the direction of the straight line connecting the two particles. This situation is represented graphically in Figure 1 (b), where the test particle is at A, the retarded source particle is at C, and the segment AC has null length. The mismatch between the two directions is very puzzling.

When the interaction is time symmetric, the retarded and the advanced four-forces could point towards their corresponding source particles, and thus have the correct intuitive direction. By postulate (Postulate I) we assume that this indeed happens [3]. As seen from Figure 1 (c), due to the symmetry of this configuration under a time inversion operation, the retarded $\mathbf{F}_{ret}$ and the advanced $\mathbf{F}_{adv}$ four-forces must be the mirror image of each other.

$$\mathbf{F}_{ret} = \left(\frac{Qq}{2R^2}, 0, 0, i\frac{Qq}{2R^2} \right) \tag{1}$$

$$\mathbf{F}_{adv} = \left(\frac{Qq}{2R^2}, 0, 0, -i\frac{Qq}{2R^2} \right) \tag{2}$$

The spatial (real) components add together, but the temporal (imaginary) components cancel each other. The resultant electrostatic four-force

$$\mathbf{F} = \mathbf{F}_{ret} + \mathbf{F}_{adv} = \left(\frac{Qq}{R^2}, 0, 0, 0 \right) \tag{3}$$

matches the standard expression. The total four-force $\mathbf{F}$ is orthogonal to the four-velocity of the test particle, even though the individual retarded and advanced four-forces are not.

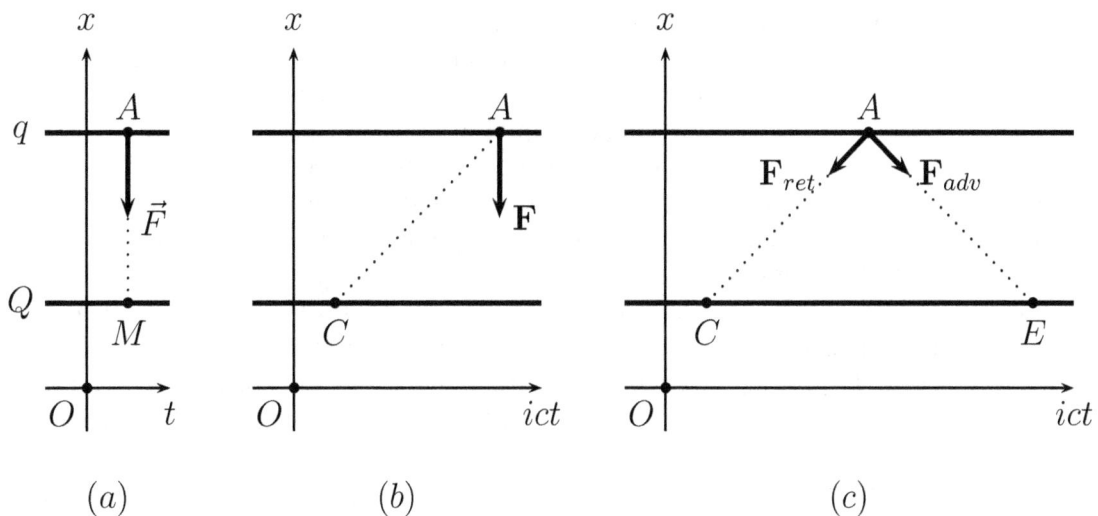

Figure 1: Examples of (a) instantaneous, (b) retarded, and (c) time symmetric electrostatic interactions.

This situation is represented graphically in Figure 1 (c), where the test particle is at A, the retarded source particle is at C, and the advanced source particle is at E.

We also notice that, in this very simple case when both particles are at rest, the retarded four-force with which the particle at C acts on the particle at A is equal in magnitude but opposite in direction to the advanced four-force with which the particle at A acts on the particle at C. The same holds true for the interaction between the particles at A and at E.

Several other authors have also considered this type of interaction, where the four-forces (the exchanges in four-momentum) are directed along the null segments connecting the interacting particles. Gen Yoneda [4] states that "it is natural to assume" that the retarded and advanced four-forces have the same direction as the displacement four-vectors connecting the interacting particles, and names this "the parallel condition". Fokker [5] postulates that "the mass vector of the transferred fragment is directed along the zero interval, so that the mass of the fragment is equal to zero." Synge [6] is even more explicit: "At the event P some entity possessing momentum and energy (i.e. a momentum-energy 4-vector) leaves the first particle and travels in a straight line with the fundamental velocity c (i.e. its world line is a straight null line) to meet the other particle at the event Q, where it is absorbed." This entity, possessing energy E and linear momentum E/c, is like a real photon, with one difference: the energy E is positive for a repulsive force, but negative for an attractive force. This classical counterpart of a quantum mechanical exchange particle is called "telehapsis" by Fokker and "internal impulse" by Synge. During their interaction in spacetime the two particles are exchanging a continuous stream of internal impulses, and it is quite remarkable the fact that "for particles with constant proper masses and relative velocity small compared with c, the inverse square law is a consequence of the mechanical laws of conservation"[7]. Due to the conservation of total four-momentum whenever the internal impulses are emitted or absorbed, this interaction mechanism "embodies the simplest available generalization to

21

relativity of Newton's law of action and reaction"[6].

As noticed by Synge, this model of interaction must be time symmetric. Since only the emission (or only the absorption) of a single real photon cannot leave the rest mass of the test particle invariant (due to the triangle inequality in Minkowski space, applied to four-momenta), in order for this rest mass to remain constant we need to have a simultaneous exchange of internal impulses with both the retarded and the advanced source particles.

The assumed time symmetry of the interaction has important physical and philosophical implications. For those investigating the nature and ontology of spacetime, as emphasized by Vesselin Petkov, "the main question is whether the world is three-dimensional or four-dimensional"[8]. Only the four-dimensionalist view is compatible with time symmetric interactions. Once the growing block universe theory is rejected and eternalism is embraced, we conclude that "unlike the pre-relativistic division of events, the relativistic division does not affect the existence of the events – the events in the past light cone, the event O, and the events in the future light cone are all *equally* existent"[8] and that "if we consider the worldline of a particle in spacetime, the worldline is a monolithic four-dimensional entity which exists timelessly in the frozen world of Minkowski spacetime"[8]. From the point of view of *classical physics* the future is already there and cannot be changed. Only by introducing *quantum mechanics* into the theory one can hope to address complex topics such as the flow of time, the human conscience, the animal conscience, free will, and acausal change.

2.2 Test particle with only radial velocity $\vec{v} = (v_x, 0, 0)$

When the test particle has a purely radial velocity $\vec{v} = (v_x, 0, 0)$, that means when the velocity has the same radial direction as the displacement vector $\vec{R} = (R, 0, 0)$, the electric field $\vec{E} = (Q/R^2, 0, 0)$, and the electrostatic force $\vec{F} = (Qq/R^2, 0, 0)$, the four-force becomes

$$\mathbf{F} = \left(\gamma \vec{F}, i\frac{\gamma}{c} \vec{F} \cdot \vec{v} \right) = \left(\gamma \frac{Qq}{R^2}, 0, 0, i\gamma \frac{v_x}{c} \frac{Qq}{R^2} \right), \tag{4}$$

where $\gamma = 1/\sqrt{1 - v^2/c^2}$ is the Lorentz factor and $v = |v_x|$ is the speed.

We notice that the four-force depends explicitly on the velocity of the test particle. "But, from a *geometrical* point of view, a point in Minkowski space is just a fixed point – it does not have a velocity!"[3] The four-force, through the distance R, also depends implicitly on the velocity of the source particle, since this distance is measured in the inertial reference frame where the source particle is instantaneously at rest. If given only two spacetime points connected by a segment of null length, it is impossible to find a relativistically invariant non-zero distance. How can we explain this dependence of the four-force on the velocities of the interacting particles? We have to acknowleddge the fact that the material point particle model, perfectly valid in the 3D Euclidean space, looses its applicability in the 4D Minkowski space. Here, in spacetime, the interaction takes place not between points, but between segments of infinitesimal length along the worldlines of the particles. Elementary particles cannot have a null dimension along the time axis of their proper reference frame. While exploring the concept of inertia, Kevin Brown has also reached the conclusion that "even an object with zero spatial extent has non-zero temporal extent"[9].

A theory of time symmetric action-at-a-distance electrodynamic interactions between length elements along the worldlines of the particles has been proposed by Fokker [10] as early as 1929. The corresponding effective segments of infinitesimal length ("entsprechenden effektiven Elementen"), who have the important property that their end points are connected by light signals, enter the expression of the Fokker action in a totally symmetric manner, with no distinction between source and test particles. This leads to the conservation of four-momentum, most clearly stated by Wheeler and Feynman: "The impulse communicated to a over the portion $d\alpha$ of its world line via retarded forces, for example, from the stretch $d\beta$ of the world line of b is equal in magnitude and opposite in sign to the impulse transfer from a to b via advanced forces over the same world line intervals (*equality of action and reaction*)." [11] Their algebraic reformulation of Fokker's theory, making use of Dirac delta functions, was demonstrated to be equivalent to classical electrodynamics. Unfortunately, since Wheeler and Feynman have assumed that the Dirac delta function is non-zero in just one point, the geometrical insight that we are looking for was lost. In order to safeguard our geometrical intuition, we assume instead that the Dirac delta function is non-zero inside an infinitesimal interval [12, 13]. Another important observation is that, unlike in our theory, Fokker [10] and Wheeler and Feynman [11] do not implement "the parallel condition", but keep each of the individual retarded and advanced four-forces orthogonal to the four-velocity of the particle on which they act uppon.

Surprisingly, even Hermann Minkowski was thinking about this alternative way of describing the interaction between two particles, seen as worldline segments of infinitesimal length whose endpoints are connected by light signals, as demonstrated by the words that he wrote when presenting his formula for the gravitational four-force. "Imagine the spacetime threads of F and F^* with the main lines in them. Let us take an infinitely small element BC on the main line of F, further on the main line of F^*, B^* is the point light source of B and C^* is the point light source of C;"[1]

Another way to proclaim that the interaction does not take place between points in Minkowski space, but between corresponding length elements, is to reffer to the infinitesimal but non-zero "thickness of the light cone".

Hugo Tetrode [14] mentions that only the infinitesimal neighborhood of the lightcone ("daß nur die infinitesimale Umgebung des Lichtkegels") with vertex at the field point contributes to the electromagnetic potentials.

Nicholas Wheeler [15] draws a lightcone with "some small but finite" thickness and writes: "If the lightcone had 'thickness' then the presence of the Doppler factor in (456) could be understood qualitatively to result from the relatively 'longer look' that the field point gets at approaching charges, the relatively 'briefer look' at receding charges."

Kevin Brown [9] also draws a diagram in which "the light cone is shown with a non-zero thickness to illustrate that the duration of time spent by each particle as it passes through the light cone depends on the speed of the particle."

The key observation is that, for two infinitesimal spacetime segments whose endpoints are connected by light signals, the ratio of their lengths depends on their orientation. For two particles at rest, the ratio is equal to one. However, when one particle is at rest, but the other particle is in motion, the ratio is no longer equal to one. These two situations are represented graphically in Figures 2 and 3. We also notice that this ratio is a Lorentz invariant. What really matters is the relative velocity.

The exact formula for the ratio of the lengths of the corresponding infinitesimal segments

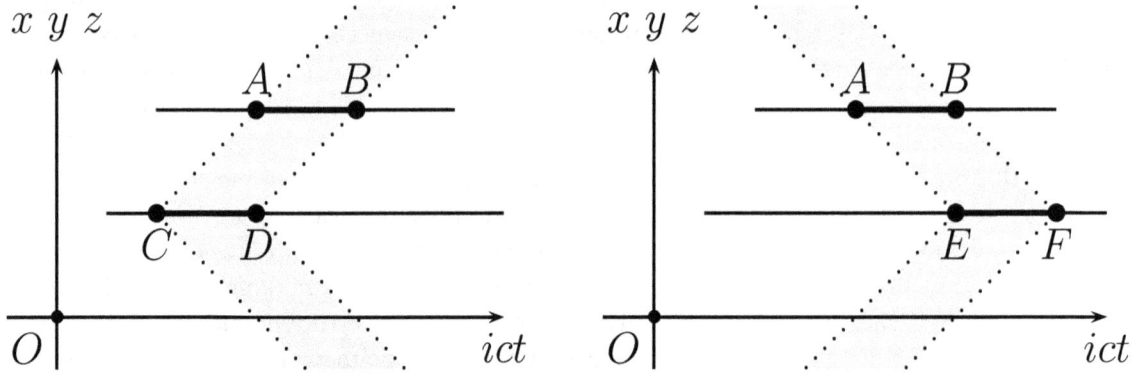

Figure 2: A test particle AB, at rest relative to the source particle, feels a retarded interaction from CD and an advanced interaction from EF.

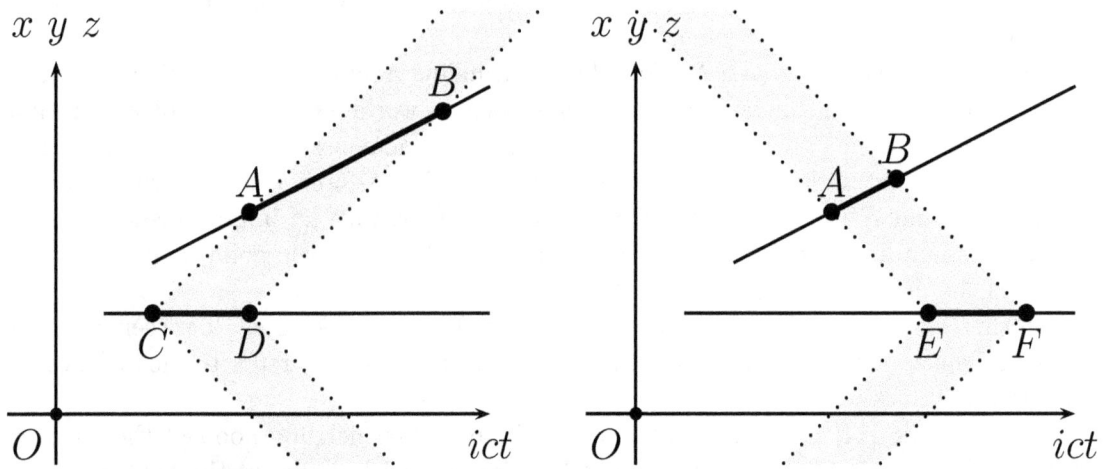

Figure 3: A test particle AB, in motion relative to the source particle, feels a retarded interaction from CD and an advanced interaction from EF.

is derived in Appendix I. In general we have

$$\frac{AB}{CD} = \frac{(\mathbf{X}_2 - \mathbf{X}_{1ret}) \cdot \mathbf{V}_{1ret}}{(\mathbf{X}_2 - \mathbf{X}_{1ret}) \cdot \mathbf{V}_2}, \tag{5}$$

$$\frac{AB}{EF} = \frac{(\mathbf{X}_2 - \mathbf{X}_{1adv}) \cdot \mathbf{V}_{1adv}}{(\mathbf{X}_2 - \mathbf{X}_{1adv}) \cdot \mathbf{V}_2}, \tag{6}$$

where $\mathbf{X}_{1ret}$ is the position four-vector of the retarded source particle CD, $\mathbf{X}_{1adv}$ is the position four-vector of the advanced source particle EF, $\mathbf{X}_2$ is the position four-vector of the test particle AB, $\mathbf{V}_{1ret}$ is the four-velocity of the retarded source particle, $\mathbf{V}_{1adv}$ is the four-velocity of the advanced source particle, and $\mathbf{V}_2$ is the four-velocity of the test particle.

24

We work in the proper reference frame of the source particle at rest. The four-velocity of the source particle (indexed with subscript 1) is

$$\mathbf{V}_{1ret} = \mathbf{V}_{1adv} = (\vec{0}, ic) = (0, 0, 0, ic). \tag{7}$$

The four velocity of the test particle (indexed with subscript 2), moving with radial velocity v_x along the x axis, is

$$\mathbf{V}_2 = (\gamma \vec{v}, i\gamma c) = (\gamma v_x, 0, 0, i\gamma c). \tag{8}$$

For the retarded interaction, the displacement four-vector is

$$\mathbf{X}_2 - \mathbf{X}_{1ret} = (\vec{R}, iR) = (R, 0, 0, iR). \tag{9}$$

For the advanced interaction, the displacement four-vector is

$$\mathbf{X}_2 - \mathbf{X}_{1adv} = (\vec{R}, -iR) = (R, 0, 0, -iR). \tag{10}$$

The ratios of the corresponding segments are calculated according to formulas (5) and (6). For the retarded interaction

$$\frac{AB}{CD} = \frac{(R, 0, 0, iR) \cdot (0, 0, 0, ic)}{(R, 0, 0, iR) \cdot (\gamma v_x, 0, 0, i\gamma c)} = \frac{-Rc}{\gamma R v_x - \gamma R c} = \frac{1}{\gamma (1 - v_x/c)}, \tag{11}$$

and for the advanced interaction

$$\frac{AB}{EF} = \frac{(R, 0, 0, -iR) \cdot (0, 0, 0, ic)}{(R, 0, 0, -iR) \cdot (\gamma v_x, 0, 0, i\gamma c)} = \frac{Rc}{\gamma R v_x + \gamma R c} = \frac{1}{\gamma (1 + v_x/c)}. \tag{12}$$

To make further progress with our theory of time symmetric action-at-a-distance electrodynamics, we recall an observation made by Olivier Costa de Beauregard [16]. The equations of motion of a relativistic point particle and the equations describing a classical elastic string in static equilibrium are isomorphic (of similar form and structure). We can think of the worldline of a particle as an elastic string at rest in spacetime. This isomorphism was also independently rediscovered in Ref. [3].

A relativistic point particle with four-momentum $\mathbf{P}$, subject to a four-force $\mathbf{F}$, undergoes a trajectory described by

$$\mathbf{P}_B - \mathbf{P}_A = \mathbf{F} \, d\tau, \tag{13}$$

where A and B are two infinitesimally close points on the worldline of the particle, separated by a proper time interval $d\tau$. The four-momentum $\mathbf{P}$ is a time-like vector directed towards the future.

A static classical elastic string under a tension of magnitude T, and subject to a linear force density $\mathbf{f}$, has segments of infinitesimal length ds in static equilibrium. We give the tension vector $\mathbf{T}$ the direction that points towards the future. The vector $\mathbf{T}$ is tangent to the string. The static equilibrium condition for the string segment AB of length ds becomes

$$\mathbf{T}_B + (-\mathbf{T}_A) + \mathbf{f} \, ds = 0, \tag{14}$$

where the minus sign indicates that the tension forces acting at the end points of the infinitesimal segment AB have opposite directions.

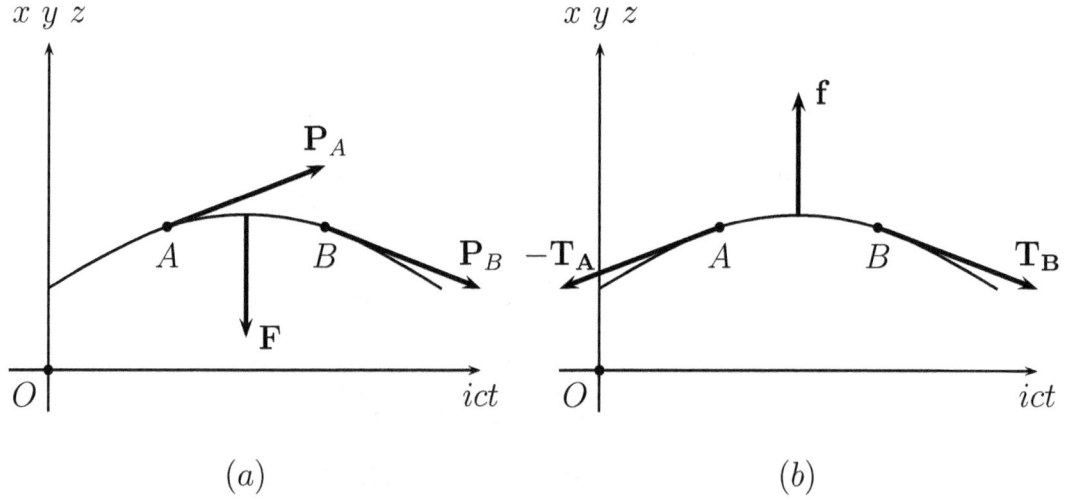

Figure 4: The worldline of a relativistic point particle (a) can be looked upon as a spacetime string in static equilibrium (b).

Equations (13) and (14) are equivalent, provided that the length of the worldline segment AB is $ds = ic\,d\tau$, the tension in the string is $\mathbf{T} = \mathbf{P}/\tau_o$, and the linear force density is $\mathbf{f} = -\mathbf{F}/s_o$. Here τ_o is a yet undetermined real constant quantity with units of time, introduced for dimensional reasons, and $s_o = ic\tau_o$. We notice that the four-force acting on the material point particle and the linear four-force density acting on the infinitesimal worldline segment have opposite directions, as seen from Fig. 4.

According to the spacetime string model described so far, when both electrically charged particles are at rest, as considered in Subsection 2.1, the retarded and the advanced linear four-force densities are

$$\mathbf{f}_{ret} = \frac{-1}{s_o}\mathbf{F}_{ret} = \frac{-1}{s_o}\Big(\frac{Qq}{2R^2}, 0, 0, i\frac{Qq}{2R^2}\Big), \tag{15}$$

$$\mathbf{f}_{adv} = \frac{-1}{s_o}\mathbf{F}_{adv} = \frac{-1}{s_o}\Big(\frac{Qq}{2R^2}, 0, 0, -i\frac{Qq}{2R^2}\Big). \tag{16}$$

We are now ready to describe, in a geometrical manner, the velocity dependence of the four-force (4). By postulate (Postulate II) we assume that, in the inertial reference frame where the source particle is instantaneously at rest, the linear four-force density acting on the test particle is proportional to the product of the Coulombian electrostatic force with the ratio of the lenght of the infinitesimal segment on the worldline of the test particle to the length of the corresponding infinitesimal segment on the worldline of the source particle [17].

For a test particle at rest, $AB/CD = 1$ and $AB/EF = 1$, which means that equations (15)-(16), and therefore also equations (1)-(3), stay the same.

However, for a test particle with only radial velocity, the retarded and the advanced

26

linear four-force densities become

$$\mathbf{f}_{ret} = \frac{-1}{s_o}\left(\frac{Qq}{2R^2}, 0, 0, i\frac{Qq}{2R^2}\right)\frac{AB}{CD}$$

$$= \frac{-1}{s_o}\left(\frac{Qq}{2R^2}\frac{1}{\gamma\left(1 - v_x/c\right)}, 0, 0, i\frac{Qq}{2R^2}\frac{1}{\gamma\left(1 - v_x/c\right)}\right), \quad (17)$$

$$\mathbf{f}_{adv} = \frac{-1}{s_o}\left(\frac{Qq}{2R^2}, 0, 0, -i\frac{Qq}{2R^2}\right)\frac{AB}{EF}$$

$$= \frac{-1}{s_o}\left(\frac{Qq}{2R^2}\frac{1}{\gamma\left(1 + v_x/c\right)}, 0, 0, -i\frac{Qq}{2R^2}\frac{1}{\gamma\left(1 + v_x/c\right)}\right). \quad (18)$$

Due to the "magic" fact that

$$\frac{1}{(1 - v_x/c)} + \frac{1}{(1 + v_x/c)} = \frac{2}{1 - v_x^2/c^2} = 2\gamma^2, \quad (19)$$

$$\frac{1}{(1 - v_x/c)} - \frac{1}{(1 + v_x/c)} = \frac{2v_x/c}{1 - v_x^2/c^2} = 2\gamma^2\frac{v_x}{c}, \quad (20)$$

the total linear four-force density becomes

$$\mathbf{f} = \mathbf{f}_{ret} + \mathbf{f}_{adv} = \frac{-1}{s_o}\left(\frac{Qq}{R^2}\gamma, 0, 0, i\frac{Qq}{R^2}\gamma\frac{v_x}{c}\right), \quad (21)$$

and the total four-force $\mathbf{F} = -s_0\,\mathbf{f}$ is in full agreement with equation (4).

Please keep in mind that equations (19)-(20) hold only when the velocity of the test particle is purely radial, that means parallel to the electric field of the stationary source particle.

A geometrical derivation of formulas (11) and (12), for the case of only radial motion, was given in Ref. [3]. Here we give a different geometrical derivation, as a preliminary step towards a discussion of the relationship between the four-forces of action and reaction. Since under a time reversal operation the velocity $\vec{v}$ of the test particle changes sign, and the formulas (11) and (12) turn into each other, it is enough to demonstrate just one of them. In particular, we assume that $v_x > 0$ and we calculate the ratio AB/CD associated with the retarded interaction. Due to the purely radial motion we can also assume that the origin of the proper reference frame K of the source particle and the origin of the proper reference frame K' of the test particle coincide when $t = t' = 0$. Each electrically charged particle is at rest at the origin of its proper reference frame. Their worldlines are the ict and ict' time axes. This situation is represented graphically in Fig. 5.

From point A we draw a line perpendicular to the ict axis, which intersects it at point T. The length of segment AT is R. From point C we draw a line perpendicular to the ict' axis, which intersects it at point U. The length of segment CU is R'. Segment AC has a null length. As a consequence, the length of segment CT is iR and the length of segment UA is iR'. Since segments CA and DB are parallel, due to the theorem of Thales we have

$$\frac{AB}{CD} = \frac{OA}{OC}. \quad (22)$$

27

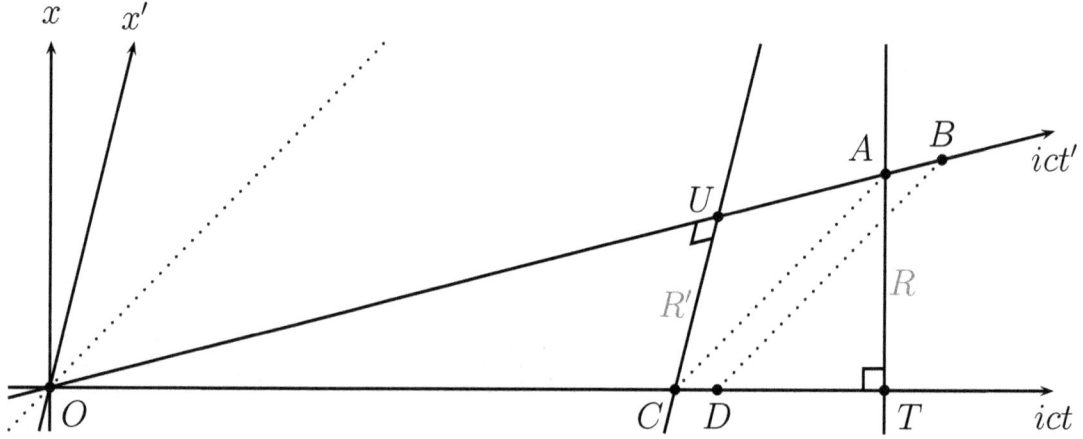

Figure 5: When the relative velocity is along the radial direction, the interacting particles AB and CD are in the same Minkowski plane.

Let t be the time coordinate of point A in the proper reference frame of the source charge. It follows that

$$AT = v_x t, \tag{23}$$

$$OT = ict, \tag{24}$$

$$OA = \sqrt{OT^2 + AT^2} = i\sqrt{c^2 - v_x^2}\, t, \tag{25}$$

$$CT = i\, AT = iv_x t, \tag{26}$$

$$OC = OT - CT = ict - iv_x t, \tag{27}$$

$$\frac{AB}{CD} = \frac{OA}{OC} = \frac{i\sqrt{c^2 - v_x^2}\, t}{ict - iv_x t} = \frac{\sqrt{c^2 - v_x^2}}{c - v_x} = \frac{1}{\gamma\left(1 - v_x/c\right)}, \tag{28}$$

and this concludes the geometrical proof of formula (11).

We want to verify that the retarded four-force acting on segment AB, due to segment CD, has the same magnitude and opposite direction as the advanced four-force acting on segment CD, due to segment AB. We find each four-force by multiplying the relevant linear four-force density by the length of the worldline segment on which it acts. According to the mechanism described, in the proper reference frame K of particle CD the retarded four-force acting on particle AB is

$$\mathbf{f}_{ret}\, AB = \frac{-1}{s_o}\left(\frac{Qq}{2R^2}\frac{AB}{CD}AB, 0, 0, i\frac{Qq}{2R^2}\frac{AB}{CD}AB\right). \tag{29}$$

Similarly, in the proper reference frame K' of particle AB the advanced four-force acting on particle CD is

$$\mathbf{g}_{adv}\, CD = \frac{-1}{s_o}\left(-\frac{Qq}{2R'^2}\frac{CD}{AB}CD, 0, 0, -i\frac{Qq}{2R'^2}\frac{CD}{AB}CD\right). \tag{30}$$

28

The easiest way to convert the four-force components from K' to K is to notice that in reference frame K'

$$\overrightarrow{CA} = \overrightarrow{CU} + \overrightarrow{UA} = (R', 0, 0, 0) + (0, 0, 0, iR') = (R', 0, 0, iR'), \tag{31}$$

while in reference frame K

$$\overrightarrow{CA} = \overrightarrow{CT} + \overrightarrow{TA} = (0, 0, 0, iR) + (R, 0, 0, 0) = (R, 0, 0, iR). \tag{32}$$

The components in K of such a vector of null length are equal to the components in K' multiplied by R/R'. Accordingly, in the proper reference frame K of particle CD the advanced four-force acting on particle CD is

$$\mathbf{g}_{adv}\, CD = \frac{-1}{s_o}\left(-\frac{Qq}{2R'^2}\frac{CD}{AB}CD\frac{R}{R'}, 0, 0, -i\frac{Qq}{2R'^2}\frac{CD}{AB}CD\frac{R}{R'}\right). \tag{33}$$

Comparing the four-force expressions (29) and (33), we see that the principle of action and reaction is verified if and only if

$$\frac{1}{R^2}\frac{AB}{CD}AB = \frac{1}{R'^2}\frac{CD}{AB}CD\frac{R}{R'}, \tag{34}$$

an equation that simplifies to

$$\frac{AB}{CD} = \frac{R}{R'}. \tag{35}$$

Due to equation (22), condition (35) is equivalent to

$$\frac{OA}{OC} = \frac{AT}{CU}, \tag{36}$$

an equation that follows from the similarity of $\triangle OTA$ and $\triangle OUC$. An algebraic derivation of equation (35) is given in Appendix B. We also notice that equation (35) is related to the surface area of $ABDC$, which, up to the first order in the infinitesimals, is equal to $AB \times R' = CD \times R$.

As a side note, as seen from equation (34), Postulate II together with the principle of action and reaction can explain why the electrostatic force has an inverse square dependence on the distance between charges.

2.3   Test particle with non-radial velocity $\vec{v} = (v_x, v_y, 0)$

When the test particle has a non-radial velocity $\vec{v} = (v_x, v_y, 0)$, the speed is $v = \sqrt{v_x^2 + v_y^2}$ and the Lorentz factor is $\gamma = 1/\sqrt{1 - (v_x^2 + v_y^2)/c^2}$.

With this new value of the Lorentz factor, the four-force has the same expression (4), and the ratios of corresponding segments have the same expressions (11) and (12). A geometrical derivation of formulas (11) and (12), for the case of non-radial motion, was also given in Ref. [3].

In addition to the algebraic derivation from Appendix B, we give here an alternative geometrical proof of the fact that, for the retarded interaction, the value of R/R' is equal to the value of AB/CD from equation (11).

We select a reference frame in which the retarded source particle at C is at rest at the origin. At time t the test particle is at point A on the Ox axis. The Ox axis represents the radial direction. The origin on the time axis is chosen in such a way that, at the initial time $t_0 = 0$, the test particle is at point W on the Oy axis. This situation is represented graphically in Fig. 6.

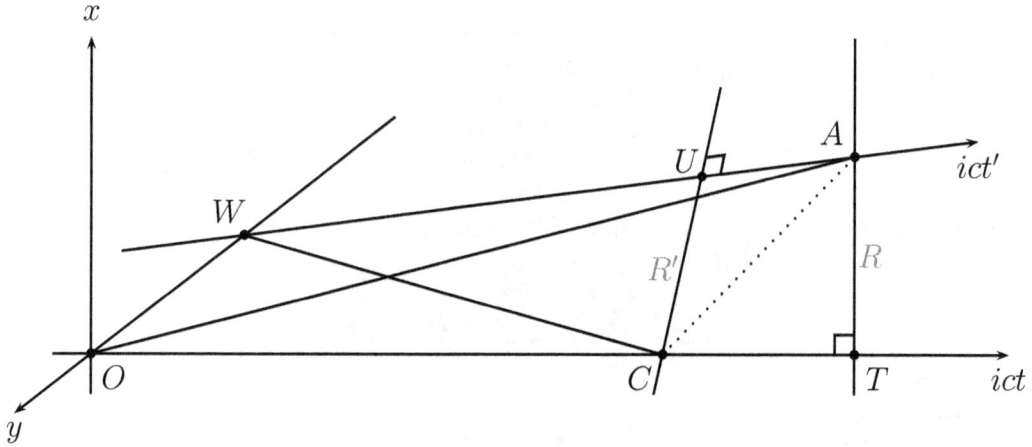

Figure 6: When the relative velocity is not along the radial direction, the worldlines of the two interacting particles, the ict and ict' time axes, are no longer in the same Minkowski plane.

It follows that

$$AT = v_x\, t = R, \tag{37}$$

$$OW = v_y\, t, \tag{38}$$

$$OT = ict, \tag{39}$$

$$OA = \sqrt{AT^2 + OT^2} = \sqrt{v_x^2 - c^2}\, t = i\sqrt{c^2 - v_x^2}\, t, \tag{40}$$

$$CT = i\, AT = iv_x\, t, \tag{41}$$

$$OC = OT - CT = ict - iv_x\, t = i(c - v_x)t, \tag{42}$$

$$WC = \sqrt{OW^2 + OC^2} = \sqrt{v_y^2 - (c - v_x)^2}\, t = i\sqrt{(c - v_x)^2 - v_y^2}\, t, \tag{43}$$

$$WA = \sqrt{OW^2 + OA^2} = \sqrt{v_y^2 + v_x^2 - c^2}\, t = i\sqrt{c^2 - v_x^2 - v_y^2}\, t, \tag{44}$$

$$CU = R', \tag{45}$$

$$UA = i\, CU = iR', \tag{46}$$

$$WU = WA - UA = i\sqrt{c^2 - v_x^2 - v_y^2}\, t - iR'. \tag{47}$$

In order to find an expression for R' we write the Pythagorean theorem in $\triangle WUC$

$$WC^2 = WU^2 + CU^2. \tag{48}$$

We solve the resulting equation

$$\left(i\sqrt{(c-v_x)^2-v_y^2}\,t\right)^2 = \left(i\sqrt{c^2-v_x^2-v_y^2}\,t - iR'\right)^2 + (R')^2, \qquad (49)$$

and we find, as expected, that

$$R' = \frac{v_x\,(c-v_x)t}{\sqrt{c^2-v_x^2-v_y^2}} = \frac{c-v_x}{\sqrt{c^2-v_x^2-v_y^2}}R = \gamma\left(1-\frac{v_x}{c}\right)R. \qquad (50)$$

As an added benefit, this geometrical proof of the formula for the R/R' ratio also allows us to discover the surprising fact that $\triangle WAC$ and $\triangle OAC$ have the same area. Indeed

$$WA \times CU = i\sqrt{c^2-v_x^2-v_y^2}\,t \times \frac{v_x\,(c-v_x)t}{\sqrt{c^2-v_x^2-v_y^2}} = OC \times AT. \qquad (51)$$

How can this be, since $\triangle OAC$ is the projection onto plane $xOict$ of $\triangle WAC$? The well known results from Euclidean geometry do not apply here, in Minkowski space, because in this specific example the two Minkowski planes intersect along a null line.

Due to the new value of the Lorentz factor, the last equalities in equations (19) and (20) no longer hold. When adding equations (17) and (18), instead of equation (21) we now obtain

$$\mathbf{f} = \mathbf{f}_{ret} + \mathbf{f}_{adv} = \frac{-1}{s_o}\left(\frac{Qq}{R^2}\gamma, 0, 0, i\frac{Qq}{R^2}\gamma\frac{v_x}{c}\right)\frac{1}{\gamma^2\left(1-v_x^2/c^2\right)}. \qquad (52)$$

We notice that there is an extra factor

$$\Gamma = \frac{1}{\gamma^2\left(1-v_x^2/c^2\right)} = \frac{c^2-v^2}{c^2-v_x^2} = \frac{c^2-v_x^2-v_y^2}{c^2-v_x^2} \qquad (53)$$

that makes our expression of the four-force a little bit different from what we expect from classical electrodynamics. Formula (53) shows that only v_y, the non-radial component of the velocity of the test particle, perpendicular to the radial electric field $\vec{E} = (E_x, 0, 0)$ of the source particle at rest, can make the Γ factor deviate from a unit value.

From an experimental point of view, the velocity of electrons in metals is not relativistic, and the correction due to the Γ factor is very small. In linear particle accelerators the velocity of the electric charges is parallel to the electric field, and this is just like the case of motion in the radial direction. In synchrotons relativistic electrons or protons are held in circular orbits not by electrostatic fields, but by magnetic fields.

From a theoretical point of view, as was noticed by Gen Yoneda in Appendix A of Ref. [4], we expect to see some disagreements between classical electrodynamics and an alternative theory that implements "the parallel condition".

2.4 Test particle with non-radial velocity $\vec{v} = (v_x, v_y, v_z)$

Consider a source particle with electric charge Q at rest at the origin $(0, 0, 0)$ of the reference frame, and a test particle with electric charge q at the position given by $\vec{R} = (R_x, R_y, R_z)$.

The distance between the two particles is $R = \sqrt{R_x^2 + R_y^2 + R_z^2}$. The test particle has velocity $\vec{v} = (v_x, v_y, v_z)$ and the Lorentz factor is $\gamma = 1/\sqrt{1 - v^2/c^2} = 1/\sqrt{1 - (v_x^2 + v_y^2 + v_z^2)/c^2}$.

The ratios of the corresponding segments are calculated according to formulas (5) and (6). For the retarded interaction

$$\frac{AB}{CD} = \frac{(\vec{R}, iR) \cdot (\vec{0}, ic)}{(\vec{R}, iR) \cdot (\gamma \vec{v}, i\gamma c)} = \frac{-Rc}{\gamma \vec{R} \cdot \vec{v} - \gamma Rc} = \frac{1}{\gamma (1 - v_{rad}/c)}, \tag{54}$$

and for the advanced interaction

$$\frac{AB}{EF} = \frac{(\vec{R}, -iR) \cdot (\vec{0}, ic)}{(\vec{R}, -iR) \cdot (\gamma \vec{v}, i\gamma c)} = \frac{Rc}{\gamma \vec{R} \cdot \vec{v} + \gamma Rc} = \frac{1}{\gamma (1 + v_{rad}/c)}, \tag{55}$$

where by definition the radial component of the velocity is $v_{rad} = \vec{v} \cdot \vec{R}/R$. We also have a non-radial component, such that $v^2 = v_{rad}^2 + v_{nonrad}^2$.

The retarded and the advanced linear four-force densities become

$$\mathbf{f}_{ret} = \frac{-1}{s_o} \left(\frac{Qq}{2R^2} \frac{\vec{R}}{R} \frac{1}{\gamma (1 - v_{rad}/c)}, i \frac{Qq}{2R^2} \frac{1}{\gamma (1 - v_{rad}/c)} \right), \tag{56}$$

$$\mathbf{f}_{adv} = \frac{-1}{s_o} \left(\frac{Qq}{2R^2} \frac{\vec{R}}{R} \frac{1}{\gamma (1 + v_{rad}/c)}, -i \frac{Qq}{2R^2} \frac{1}{\gamma (1 + v_{rad}/c)} \right), \tag{57}$$

and the total linear four-force density becomes

$$\mathbf{f} = \mathbf{f}_{ret} + \mathbf{f}_{adv} = \frac{-1}{s_o} \left(\frac{Qq}{R^2} \frac{\vec{R}}{R} \gamma, i \frac{Qq}{R^2} \gamma \frac{v_{rad}}{c} \right) \frac{1}{\gamma^2 (1 - v_{rad}^2/c^2)}. \tag{58}$$

We notice that there is an extra factor

$$\Gamma = \frac{1}{\gamma^2 (1 - v_{rad}^2/c^2)} = \frac{c^2 - v^2}{c^2 - v_{rad}^2} = \frac{c^2 - v_{rad}^2 - v_{nonrad}^2}{c^2 - v_{rad}^2} \tag{59}$$

that makes our expression of the four-force

$$\mathbf{F} = -s_0 \mathbf{f} = \left(\frac{Qq}{R^2} \frac{\vec{R}}{R} \gamma \Gamma, i \frac{Qq}{R^2} \gamma \frac{v_{rad}}{c} \Gamma \right) \tag{60}$$

a little bit different from what we expect from classical electrodynamics. The four-force (60) is what we get when the Coulombian force $\vec{F} = Qq\vec{R}/R^3$ in equation (4) is multiplied by Γ. The four-force (60) is still orthogonal to the four-velocity of the test particle.

Although the extra Γ factor is unexpected, we cannot simply eliminate it by postulate, since that would destroy the balance between action and reaction. As proven in Appendix C, the two Γ factors for action and reaction are equal to each other only when each particle, relative to the other particle, is moving only in the radial direction.

3 Source particle in hyperbolic motion, and field point that is simultaneous with the center of the hyperbola

Consider a source particle with electric charge Q, moving in hyperbolic motion along the x axis. By our choice, the origin O of the 4D reference frame is also the center of the hyperbola. The hyperbola intersects the positive x axis in a point at a distance a from the origin. Let s be the arclength on the hyperbola, related to the proper time τ of the particle by the formula $ds = i\,c\,d\tau$. In analogy with the definition of the value in radians of an angle in 3D Euclidean space, we introduce the imaginary angle ψ based on the formula $ds = a\,d\psi$ [18]. It follows that $d\psi/d\tau = i\,c/a$. The positive direction of the angle coordinate ψ points into the future. The worldline of the source particle is described by the position four-vector

$$\mathbf{X}_1 = \Big(a\,\cos(\psi), 0, 0, a\,\sin(\psi) \Big), \tag{61}$$

and the four-velocity of the source particle is given by

$$\mathbf{V}_1 = \frac{d\mathbf{X}_1}{d\tau} = \Big(-i\,c\,\sin(\psi), 0, 0, i\,c\,\cos(\psi) \Big). \tag{62}$$

Consider a field point A that is simultaneous with the center O of the hyperbola, in a given inertial reference frame K. The position four-vector of point A is

$$\mathbf{X}_2 = \Big(\rho, y, z, 0 \Big), \tag{63}$$

Relative to field point A, there is a retarded source charge at point C and an advanced source charge at point E. The displacement four-vector from the source particle to the field point is

$$\mathbf{X}_2 - \mathbf{X}_1 = \Big(\rho - a\,\cos(\psi), y, z, -a\,\sin(\psi) \Big). \tag{64}$$

For both the retarded and the advanced electromagnetic interactions the condition $(\mathbf{X}_2 - \mathbf{X}_1) \cdot (\mathbf{X}_2 - \mathbf{X}_1) = 0$ reduces to the equation

$$\cos(\psi) = \frac{\rho^2 + y^2 + z^2 + a^2}{2a\rho}, \tag{65}$$

which has two imaginary solutions, θ and $-\theta$, that have the same absolute value. For the retarded solution $\psi_{ret} = -\theta$ is a negative imaginary angle, while for the advanced solution $\psi_{adv} = \theta$ is a positive imaginary angle.

Let M be the intersection of line CE with the x axis, as seen in Fig. 7. Since $OC = OE = a$, $\triangle OCE$ is an isosceles triangle and the angle bisector OM is also a median and a height. The **direction of the electric field** at point A, as shown in Ref. [19], is the direction of the displacement four-vector $\overrightarrow{MA}$. This direction reveals the time symmetric nature of the interaction, because [17]

$$\overrightarrow{MA} = \frac{1}{2}\overrightarrow{CA} + \frac{1}{2}\overrightarrow{EA}. \tag{66}$$

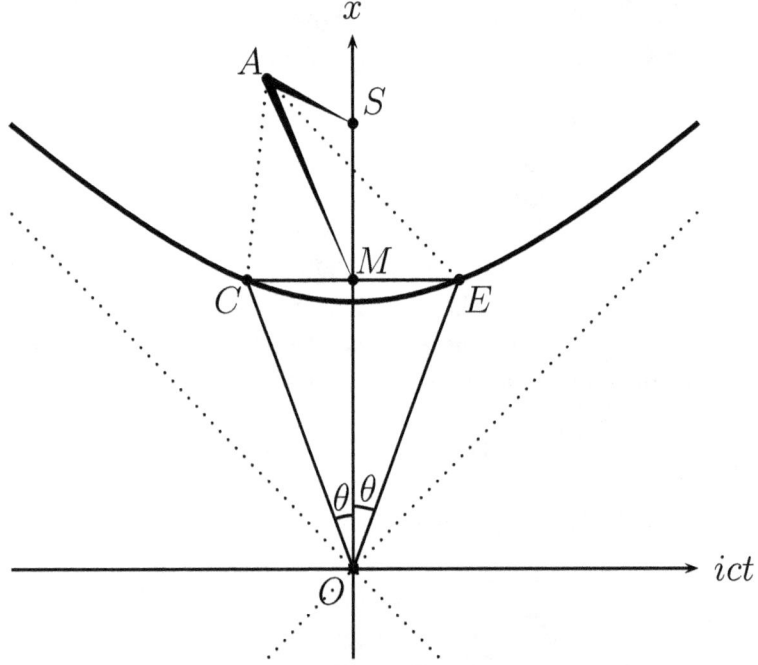

Figure 7: In the reference frame K where the field point A and the center of the hyperbola are simultaneous, the electric field has the direction of $\overrightarrow{MA}$.

Let S be the projection of point A on the x axis. Since the time axis is perpendicular to AO (by hypothesis), and since the time axis is also perpendicular to the x axis, it follows that AS is perpendicular to the time axis, which means that AS is perpendicular to the $xOict$ Minkowski plane.

We have [19]

$$OC = OE = a, \tag{67}$$

$$OM = OC\cos(\theta) = a\cos(\theta), \tag{68}$$

$$CM = ME = OC\sin(\theta) = a\sin(\theta), \tag{69}$$

$$OS = \rho, \tag{70}$$

$$AS = \sqrt{y^2 + z^2}, \tag{71}$$

$$MS = OS - OM = \rho - a\cos(\theta), \tag{72}$$

$$MA = \sqrt{MS^2 + AS^2} = \sqrt{\rho^2 + a^2\cos^2(\theta) - 2\rho a\cos(\theta) + y^2 + z^2}. \tag{73}$$

With the help of equation(65) we find that

$$MA = \sqrt{a^2\cos^2(\theta) - a^2} = a\sqrt{\cos^2(\theta) - 1} = -ia\sin(\theta). \tag{74}$$

We introduce the notation $\overrightarrow{MA} = (\vec{R}, 0)$, meaning that in reference frame K we have

$$\vec{R} = \left(\rho - a\cos(\theta), y, z\right). \tag{75}$$

34

This 3D vector with the direction of the electric field is the equivalent of the radial position vector when the source charge was at rest. The distance

$$R = -ia\sin(\theta) \tag{76}$$

is a Lorentz invariant quantity.

Through point C we draw the line tangent to the hyperbola, which intersects the x axis at point U, as seen in Fig. 8. This tangent line is the ict' time axis of the inertial reference frame K' in which the retarded source particle at C is instantaneously at rest. We also notice that the tangent line CU is perpendicular to OC, as confirmed by the equation $\mathbf{X}_1 \cdot \mathbf{V}_1 = 0$.

From point A we draw a line perpendicular to the ict' axis, which intersects it at point T. The ict' axis is perpendicular to both AT (by construction) and AS (since AS is perpendicular to any line in the $xOict$ plane), and as a result the ict' axis is perpendicular to the AST plane. As a consequence, the ict' axis is perpendicular to ST. Since the ict' axis is also perpendicular to OC, we conclude that OC is parallel to ST, and as a result the measure of $\angle UST$ is also θ.

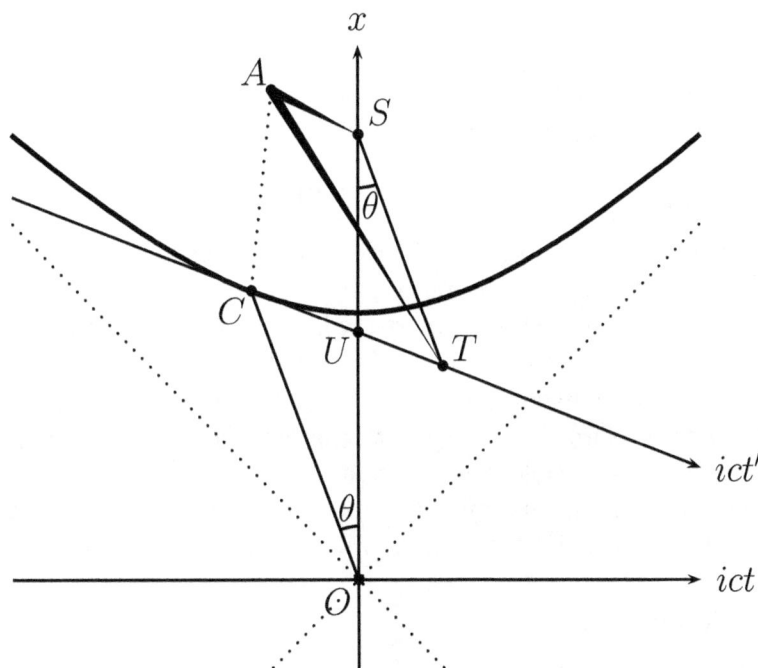

Figure 8: In the reference frame K' where the retarded source particle at C is instantaneously at rest, the position of the field point A is given by $\overrightarrow{TA}$.

We have [19]

$$OC = a, \tag{77}$$

$$OU = OC/\cos(\theta) = a/\cos(\theta), \tag{78}$$

$$OS = \rho, \tag{79}$$

$$AS = \sqrt{y^2 + z^2}, \tag{80}$$

$$US = OS - OU = \rho - a/\cos(\theta), \tag{81}$$

$$TS = US\cos(\theta) = \rho\cos(\theta) - a, \tag{82}$$

$$TA = \sqrt{TS^2 + AS^2} = \sqrt{\rho^2\cos^2(\theta) + a^2 - 2\rho a\cos(\theta) + y^2 + z^2}. \tag{83}$$

With the help of equation(65) we find that

$$TA = \sqrt{\rho^2\cos^2(\theta) - \rho^2} = \rho\sqrt{\cos^2(\theta) - 1} = -i\rho\sin(\theta). \tag{84}$$

We introduce the notation $\overrightarrow{TA} = (\vec{r}, 0)$, meaning that in reference frame K' we have

$$\vec{r} = \Big(\rho\cos(\theta) - a, y, z\Big). \tag{85}$$

The length of segment TA is the Coulombian radius r that enters the calculation of the four-force in a reference frame co-moving with the source particle. The distance

$$r = -i\rho\sin(\theta) \tag{86}$$

is a Lorentz invariant quantity. The **magnitude of the electric field** at point A, as shown in Ref. [19], is $E = Q/r^2$ (in Gaussian units).

In conclusion, in the reference frame where the field point and the center of the hyperbola are simultaneous, the particle in hyperbolic motion is the source of an electric field $\vec{E}$ with the direction of $\overrightarrow{MA}$ and the magnitude of Q/r^2. In this reference frame the magnetic field $\vec{B}$ is zero and the Lorentz force has only an electric part, that is $\vec{F} = q\,\vec{E}$.

A Lorentz transformation could bring us to a reference frame where the field point and the center of the hyperbola are not simultaneous. This case, which produces the general expression of the electromagnetic field tensor, will not be investigated here.

3.1   Test particle at rest

Consider a test particle with electric charge q, at rest at the position of field point A. We work in the reference frame K where the field point and the center of the hyperbola are simultaneous.

The four-velocity of the retarded source particle at point C is

$$\mathbf{V}_{1ret} = \Big(i\,c\,\sin(\theta), 0, 0, i\,c\,\cos(\theta)\Big). \tag{87}$$

The four-velocity of the advanced source particle at point E is

$$\mathbf{V}_{1adv} = \Big(-i\,c\,\sin(\theta), 0, 0, i\,c\,\cos(\theta)\Big). \tag{88}$$

36

The four-velocity of the test particle at rest is

$$\mathbf{V}_2 = (\vec{0}, ic) = (0, 0, 0, ic). \tag{89}$$

For the retarded interaction, the displacement four-vector is

$$\mathbf{X}_2 - \mathbf{X}_{1ret} = (\vec{R}, iR) = \left(\rho - a\cos(\theta), y, z, iR\right). \tag{90}$$

For the advanced interaction, the displacement four-vector is

$$\mathbf{X}_2 - \mathbf{X}_{1adv} = (\vec{R}, -iR) = \left(\rho - a\cos(\theta), y, z, -iR\right). \tag{91}$$

The ratios of the corresponding segments are calculated according to formulas (5) and (6). For the retarded interaction

$$\frac{AB}{CD} = \frac{(\mathbf{X}_2 - \mathbf{X}_{1ret}) \cdot \mathbf{V}_{1ret}}{(\mathbf{X}_2 - \mathbf{X}_{1ret}) \cdot \mathbf{V}_2}$$

$$= \frac{\left(\rho - a\cos(\theta), y, z, iR\right) \cdot \left(ic\sin(\theta), 0, 0, ic\cos(\theta)\right)}{\left(\rho - a\cos(\theta), y, z, iR\right) \cdot (0, 0, 0, ic)}$$

$$= \frac{ic\sin(\theta)[\rho - a\cos(\theta)] - Rc\cos(\theta)}{-Rc} = \frac{\rho}{a}, \tag{92}$$

and for the advanced interaction

$$\frac{AB}{EF} = \frac{(\mathbf{X}_2 - \mathbf{X}_{1adv}) \cdot \mathbf{V}_{1adv}}{(\mathbf{X}_2 - \mathbf{X}_{1adv}) \cdot \mathbf{V}_2}$$

$$= \frac{\left(\rho - a\cos(\theta), y, z, -iR\right) \cdot \left(-ic\sin(\theta), 0, 0, ic\cos(\theta)\right)}{\left(\rho - a\cos(\theta), y, z, -iR\right) \cdot (0, 0, 0, ic)}$$

$$= \frac{-ic\sin(\theta)[\rho - a\cos(\theta)] + Rc\cos(\theta)}{Rc} = \frac{\rho}{a}, \tag{93}$$

where we have used the fact that $i\sin(\theta) = -R/a$. We also notice that $r/R = \rho/a$, in full agreement with our conclusion from Appendix B, since r is the Coulombian radius to the test particle in the co-moving reference frame of the source particle, while R is the Coulombian radius to the (retarded or advanced) source particle in the proper reference frame of the test particle.

According to the mechanism described, in the reference frame K' that is co-moving with the retarded source particle at C, the linear four-force density acting on the particle at A is

$$\mathbf{f}_{ret} = \frac{-1}{s_o}\left(\frac{Qq}{2r^2}\frac{\rho}{a}\frac{\vec{r}}{r}, i\frac{Qq}{2r^2}\frac{\rho}{a}\right) = \frac{-1}{s_o}\frac{Qq}{2r^3}\frac{\rho}{a}(\vec{r}, ir). \tag{94}$$

What are the components of this linear four-force density in the reference frame K in which the field point and the center of the hyperbola are simultaneous? In reference frame K' we have

$$\mathbf{X}_2 - \mathbf{X}_{1ret} = \overrightarrow{CA} = \overrightarrow{CT} + \overrightarrow{TA} = (\vec{0}, ir) + (\vec{r}, 0) = (\vec{r}, ir), \tag{95}$$

while in reference frame K we have

$$\mathbf{X}_2 - \mathbf{X}_{1ret} = \overrightarrow{CA} = \overrightarrow{CM} + \overrightarrow{MA} = (\vec{0}, iR) + (\vec{R}, 0) = (\vec{R}, iR). \qquad (96)$$

As a result, in reference frame K we have

$$\mathbf{f}_{ret} = \frac{-1}{s_o} \frac{Qq}{2r^3} \frac{\rho}{a} (\vec{R}, iR) = \frac{-1}{s_o} \left(\frac{Qq}{2r^2} \frac{\rho}{a} \frac{\vec{R}}{R} \frac{R}{r}, i \frac{Qq}{2r^2} \frac{\rho}{a} \frac{R}{r} \right) = \frac{-1}{s_o} \left(\frac{Qq}{2r^2} \frac{\vec{R}}{R}, i \frac{Qq}{2r^2} \right). \qquad (97)$$

In a similar manner we derive

$$\mathbf{f}_{adv} = \frac{-1}{s_o} \left(\frac{Qq}{2r^2} \frac{\vec{R}}{R}, -i \frac{Qq}{2r^2} \right). \qquad (98)$$

The total linear four-force density becomes

$$\mathbf{f} = \mathbf{f}_{ret} + \mathbf{f}_{adv} = \frac{-1}{s_o} \left(\frac{Qq}{r^2} \frac{\vec{R}}{R}, 0 \right), \qquad (99)$$

and the total four-force $\mathbf{F} = -s_0 \mathbf{f}$ is identical to the classical result.

We notice that, in the proper reference frame of the test particle, in order for the four-force to be orthogonal to the four-velocity, we need the imaginary (temporal) components of the retarded and advanced linear four-force densities to cancel each other. This happens for a source charge in hyperbolic motion, in the reference frame in which the field point and the center of the hyperbola are simultaneous, due to the symmetry of the configuration, but this may not happen in general for a source charge in random motion.

3.2   Test particle in motion

This time the four-velocity of the test particle at the position of the field point A is

$$\mathbf{V}_2 = (\gamma_2 \, \vec{v_2}, i\gamma_2 \, c). \qquad (100)$$

The ratios of the corresponding segments are calculated according to formulas (5) and (6). Only the denominators in equations (92)-(93) are a little bit different. For the retarded interaction we have

$$(\mathbf{X}_2 - \mathbf{X}_{1ret}) \cdot \mathbf{V}_2 = (\vec{R}, iR) \cdot (\gamma_2 \, \vec{v_2}, i\gamma_2 \, c)$$
$$= \gamma_2 \vec{R} \cdot \vec{v_2} - \gamma_2 Rc = -Rc\gamma_2 \left(1 - v_{2rad}/c \right), \quad (101)$$

$$\frac{AB}{CD} = \frac{(\mathbf{X}_2 - \mathbf{X}_{1ret}) \cdot \mathbf{V}_{1ret}}{(\mathbf{X}_2 - \mathbf{X}_{1ret}) \cdot \mathbf{V}_2} = \frac{\rho}{a} \frac{1}{\gamma_2 \left(1 - v_{2rad}/c \right)}, \qquad (102)$$

and for the advanced interaction

$$(\mathbf{X}_2 - \mathbf{X}_{1adv}) \cdot \mathbf{V}_2 = (\vec{R}, -iR) \cdot (\gamma_2 \, \vec{v_2}, i\gamma_2 \, c)$$
$$= \gamma_2 \vec{R} \cdot \vec{v_2} + \gamma_2 Rc = Rc\gamma_2 \left(1 + v_{2rad}/c \right), \quad (103)$$

$$\frac{AB}{EF} = \frac{(\mathbf{X}_2 - \mathbf{X}_{1adv}) \cdot \mathbf{V}_{1adv}}{(\mathbf{X}_2 - \mathbf{X}_{1adv}) \cdot \mathbf{V}_2} = \frac{\rho}{a} \frac{1}{\gamma_2 \left(1 + v_{2rad}/c\right)}, \tag{104}$$

where by definition $v_{2rad} = \vec{v_2} \cdot \vec{R}/R$.

In reference frame K the retarded linear four-force density becomes

$$\mathbf{f}_{ret} = \frac{-1}{s_o} \left(\frac{Qq}{2r^2} \frac{1}{\gamma_2 \left(1 - v_{2rad}/c\right)} \frac{\vec{R}}{R}, i \frac{Qq}{2r^2} \frac{1}{\gamma_2 \left(1 - v_{2rad}/c\right)} \right), \tag{105}$$

and the advanced linear four-force density becomes

$$\mathbf{f}_{adv} = \frac{-1}{s_o} \left(\frac{Qq}{2r^2} \frac{1}{\gamma_2 \left(1 + v_{2rad}/c\right)} \frac{\vec{R}}{R}, -i \frac{Qq}{2r^2} \frac{1}{\gamma_2 \left(1 + v_{2rad}/c\right)} \right). \tag{106}$$

The total linear four-force density becomes

$$\mathbf{f} = \mathbf{f}_{ret} + \mathbf{f}_{adv} = \frac{-1}{s_o} \left(\gamma_2 \frac{Qq}{r^2} \frac{\vec{R}}{R}, i\gamma_2 \frac{v_{2rad}}{c} \frac{Qq}{r^2} \right) \frac{1}{\gamma_2{}^2 \left(1 - v_{2rad}^2/c^2\right)}. \tag{107}$$

The total four-force $\mathbf{F} = -s_0 \mathbf{f}$ is perpendicular to the four-velocity of the test particle, as required by the first part of equation (4).

We again notice that there is an extra factor

$$\Gamma_2 = \frac{1}{\gamma_2{}^2 \left(1 - v_{2rad}^2/c^2\right)} = \frac{c^2 - v_2{}^2}{c^2 - v_{2rad}^2} \tag{108}$$

that makes our expression of the four-force a little bit different from what we expect from classical electrodynamics.

4 The invariant expression of the four-force

We are now ready to write down the general expression of the four-force in the reference frame K that is co-moving with the test particle at point A. Since we will write this expression in an explicitly Lorentz invariant form, the expression of the four-force will be equally valid in any inertial reference frame. Let K' be the reference frame that is co-moving with the retarded source particle at point C, and let K'' be the reference frame that is co-moving with the advanced source particle at point E. As seen from Figure 9, the ict time axis is the line that goes through point A and is tangent to the worldline of the test particle, the ict' time axis is the line that goes through point C and is tangent to the worldline of the retarded source particle, and the ict'' time axis is the line that goes through point E and is tangent to the worldline of the advanced source particle.

From point A we draw a line perpendicular to the ict' axis, which intersects it at point T. The length of segment TA is R', and the length of segment CT is iR'. In reference frame K' we have $\overrightarrow{TA} = (\vec{R}', 0)$. From point C we draw a line perpendicular to the ict axis, which intersects it at point U. The length of segment UC is R_{ret}, and the length of segment UA is iR_{ret}. In reference frame K we have $\overrightarrow{CU} = (\vec{R}_{ret}, 0)$.

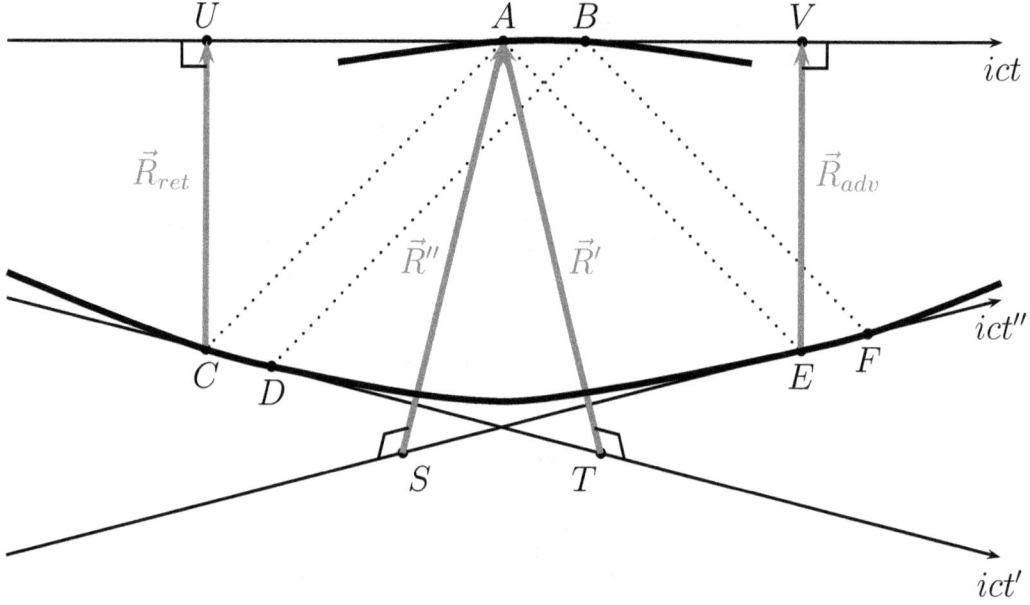

Figure 9: The test particle at AB interacts with the retarded source particle at CD and with the advanced source particle at EF. The time axes are lines tangent to the worldlines of the particles through points A, C, and E.

From point A we draw a line perpendicular to the ict'' axis, which intersects it at point S. The length of segment SA is R'', and the length of segment SE is iR''. In reference frame K'' we have $\overrightarrow{SA} = (\vec{R}'', 0)$. From point E we draw a line perpendicular to the ict axis, which intersects it at point V. The length of segment EV is R_{adv}, and the length of segment AV is iR_{adv}. In reference frame K we have $\overrightarrow{EV} = (\vec{R}_{adv}, 0)$.

In reference frame K' the retarded linear four-force density is

$$\mathbf{f}_{ret} = \frac{-1}{s_o}\left(\frac{Qq}{2R'^2}\frac{\vec{R}'}{R'}\frac{AB}{CD}, i\frac{Qq}{2R'^2}\frac{AB}{CD}\right) = \frac{-1}{s_o}\frac{Qq}{2R'^3}\frac{AB}{CD}(\vec{R}', iR'), \tag{109}$$

and the displacement four-vector $\overrightarrow{CA}$ is

$$\mathbf{X}_2 - \mathbf{X}_{1ret} = \overrightarrow{CA} = \overrightarrow{CT} + \overrightarrow{TA} = (\vec{0}, iR') + (\vec{R}', 0) = (\vec{R}', iR'). \tag{110}$$

In reference frame K the displacement four-vector $\overrightarrow{CA}$ is

$$\mathbf{X}_2 - \mathbf{X}_{1ret} = \overrightarrow{CA} = \overrightarrow{CU} + \overrightarrow{UA} = (\vec{R}_{ret}, 0) + (\vec{0}, iR_{ret}) = (\vec{R}_{ret}, iR_{ret}), \tag{111}$$

and, as a consequence, the retarded linear four-force density is

$$\mathbf{f}_{ret} = \frac{-1}{s_o}\frac{Qq}{2R'^3}\frac{AB}{CD}(\vec{R}_{ret}, iR_{ret}) = \frac{-1}{s_o}\left(\frac{Qq}{2R'^2}\frac{\vec{R}_{ret}}{R_{ret}}\frac{AB}{CD}\frac{R_{ret}}{R'}, i\frac{Qq}{2R'^2}\frac{AB}{CD}\frac{R_{ret}}{R'}\right), \tag{112}$$

which, since $R'/R_{ret} = AB/CD$ according to formula (155), simplifies to

$$\mathbf{f}_{ret} = \frac{-1}{s_o}\left(\frac{Qq}{2R'^2}\frac{\vec{R}_{ret}}{R_{ret}}, i\frac{Qq}{2R'^2}\right). \tag{113}$$

40

In reference frame K', where $\mathbf{V}_{1ret} = (\vec{0}, ic)$, we have

$$(\mathbf{X}_2 - \mathbf{X}_{1ret}) \cdot \mathbf{V}_{1ret} = (\vec{R}', iR') \cdot (\vec{0}, ic) = -R'c, \tag{114}$$

and in reference frame K, where $\mathbf{V}_2 = (\vec{0}, ic)$, we have

$$(\mathbf{X}_2 - \mathbf{X}_{1ret}) \cdot \mathbf{V}_2 = (\vec{R}_{ret}, iR_{ret}) \cdot (\vec{0}, ic) = -R_{ret}c. \tag{115}$$

The Lorentz invariant expression of the retarded four-force becomes

$$\mathbf{F}_{ret} = -s_0 \mathbf{f}_{ret} = \frac{Qq}{2R'^2 R_{ret}} (\vec{R}_{ret}, iR_{ret})$$

$$= \frac{-Qqc^3 (\mathbf{X}_2 - \mathbf{X}_{1ret})}{2[(\mathbf{X}_2 - \mathbf{X}_{1ret}) \cdot \mathbf{V}_{1ret}]^2 (\mathbf{X}_2 - \mathbf{X}_{1ret}) \cdot \mathbf{V}_2}. \tag{116}$$

In reference frame K'' the advanced linear four-force density is

$$\mathbf{f}_{adv} = \frac{-1}{s_o} \left(\frac{Qq}{2R''^2} \frac{\vec{R}''}{R''} \frac{AB}{EF}, -i \frac{Qq}{2R''^2} \frac{AB}{EF} \right) = \frac{-1}{s_o} \frac{Qq}{2R''^3} \frac{AB}{EF} (\vec{R}'', -iR''), \tag{117}$$

and the displacement four-vector $\overrightarrow{EA}$ is

$$\mathbf{X}_2 - \mathbf{X}_{1adv} = \overrightarrow{EA} = \overrightarrow{ES} + \overrightarrow{SA} = (\vec{0}, -iR'') + (\vec{R}'', 0) = (\vec{R}'', -iR''). \tag{118}$$

In reference frame K the displacement four-vector $\overrightarrow{EA}$ is

$$\mathbf{X}_2 - \mathbf{X}_{1adv} = \overrightarrow{EA} = \overrightarrow{EV} + \overrightarrow{VA} = (\vec{R}_{adv}, 0) + (\vec{0}, -iR_{adv}) = (\vec{R}_{adv}, -iR_{adv}), \tag{119}$$

and, as a consequence, the advanced linear four-force density is

$$\mathbf{f}_{adv} = \frac{-1}{s_o} \frac{Qq}{2R''^3} \frac{AB}{EF} (\vec{R}_{adv}, -iR_{adv})$$

$$= \frac{-1}{s_o} \left(\frac{Qq}{2R''^2} \frac{\vec{R}_{adv}}{R_{adv}} \frac{AB}{EF} \frac{R_{adv}}{R''}, -i \frac{Qq}{2R''^2} \frac{AB}{EF} \frac{R_{adv}}{R''} \right), \tag{120}$$

which, since $R''/R_{adv} = AB/EF$ according to formula (155), simplifies to

$$\mathbf{f}_{adv} = \frac{-1}{s_o} \left(\frac{Qq}{2R''^2} \frac{\vec{R}_{adv}}{R_{adv}}, -i \frac{Qq}{2R''^2} \right). \tag{121}$$

In reference frame K'', where $\mathbf{V}_{1adv} = (\vec{0}, ic)$, we have

$$(\mathbf{X}_2 - \mathbf{X}_{1adv}) \cdot \mathbf{V}_{1adv} = (\vec{R}'', -iR'') \cdot (\vec{0}, ic) = R''c, \tag{122}$$

and in reference frame K, where $\mathbf{V}_2 = (\vec{0}, ic)$, we have

$$(\mathbf{X}_2 - \mathbf{X}_{1adv}) \cdot \mathbf{V}_2 = (\vec{R}_{adv}, -iR_{adv}) \cdot (\vec{0}, ic) = R_{adv}c. \tag{123}$$

The Lorentz invariant expression of the advanced four-force becomes

$$\mathbf{F}_{adv} = -s_0\,\mathbf{f}_{adv} = \frac{Q\,q}{2\,R''^2\,R_{adv}}(\vec{R}_{adv}, -iR_{adv})$$

$$= \frac{Q\,q\,c^3\,(\mathbf{X}_2 - \mathbf{X}_{1adv})}{2\,[(\mathbf{X}_2 - \mathbf{X}_{1adv})\cdot\mathbf{V}_{1adv}]^2\,(\mathbf{X}_2 - \mathbf{X}_{1adv})\cdot\mathbf{V}_2}. \quad (124)$$

Adding the two contributions we obtain the total four-force

$$\mathbf{F} = \mathbf{F}_{ret} + \mathbf{F}_{adv} = \frac{Q\,q}{2\,R'^2\,R_{ret}}(\vec{R}_{ret}, iR_{ret}) + \frac{Q\,q}{2\,R''^2\,R_{adv}}(\vec{R}_{adv}, -iR_{adv})$$

$$= \frac{-Q\,q\,c^3\,(\mathbf{X}_2 - \mathbf{X}_{1ret})}{2\,[(\mathbf{X}_2 - \mathbf{X}_{1ret})\cdot\mathbf{V}_{1ret}]^2\,(\mathbf{X}_2 - \mathbf{X}_{1ret})\cdot\mathbf{V}_2}$$

$$+ \frac{Q\,q\,c^3\,(\mathbf{X}_2 - \mathbf{X}_{1adv})}{2\,[(\mathbf{X}_2 - \mathbf{X}_{1adv})\cdot\mathbf{V}_{1adv}]^2\,(\mathbf{X}_2 - \mathbf{X}_{1adv})\cdot\mathbf{V}_2}. \quad (125)$$

Our expression (125) of the total four-force is closely related to the electrostatic ("elektrostatischen") four-force of Fokker [10], which is [17]

$$\mathbf{F} = \mathbf{F}_{ret} + \mathbf{F}_{adv} = \frac{-Q\,q\,(\mathbf{V}_2 \cdot \mathbf{V}_{1ret})^2\,(\mathbf{X}_2 - \mathbf{X}_{1ret})}{2\,c\,[(\mathbf{X}_2 - \mathbf{X}_{1ret})\cdot\mathbf{V}_{1ret}]^2\,(\mathbf{X}_2 - \mathbf{X}_{1ret})\cdot\mathbf{V}_2}$$

$$+ \frac{Q\,q\,(\mathbf{V}_2 \cdot \mathbf{V}_{1adv})^2\,(\mathbf{X}_2 - \mathbf{X}_{1adv})}{2\,c\,[(\mathbf{X}_2 - \mathbf{X}_{1adv})\cdot\mathbf{V}_{1adv}]^2\,(\mathbf{X}_2 - \mathbf{X}_{1adv})\cdot\mathbf{V}_2}. \quad (126)$$

We can get formula (125) from formula (126) if we replace $(\mathbf{V}_2 \cdot \mathbf{V}_1)^2$ with $\mathbf{V}_2^2\,\mathbf{V}_1^2 = c^4$. For this reason it would be interesting to see what happens when the scalar product $\mathbf{V}_2 \cdot \mathbf{V}_1$ in Fokker's electrodynamic action [10, 11, 13] is replaced with $|\mathbf{V}_2||\mathbf{V}_1|$. Since $\mathbf{V}_2 \cdot \mathbf{V}_1 = |\mathbf{V}_2||\mathbf{V}_1|\cos(\varphi)$, where φ is the angle between the four-velocities, we are justified in making this substitution whenever the two four-velocities are parallel, or whenever the relative velocities of the two particles are very small. This condition is also assumed true in Synge's theory [6, 7]. With this substitution Fokker's action, written with a metric tensor of signature $(+, -, -, -)$

$$W_{Fokker} = -\sum_A \int m_A c \sqrt{dx_{A\alpha}\,dx_A^\alpha}$$

$$- \sum_A \sum_{B>A} \frac{q_A q_B}{c} \int \int \delta\big((x_{A\beta} - x_{B\beta})(x_A^\beta - x_B^\beta)\big)\,dx_{A\alpha}\,dx_B^\alpha, \quad (127)$$

becomes

$$W = -\sum_A \int m_A c \sqrt{dx_{A\alpha}\,dx_A^\alpha}$$

$$- \sum_A \sum_{B>A} \frac{q_A q_B}{c} \int \int \delta\big((x_{A\beta} - x_{B\beta})(x_A^\beta - x_B^\beta)\big)\,\sqrt{dx_{A\mu}\,dx_A^\mu}\,\sqrt{dx_{B\nu}\,dx_B^\nu}. \quad (128)$$

5   The principle of action and reaction

The principle of action and reaction, which has already been verified for two particles with relative motion only in the radial direction, will now be demonstrated for the most general situation. We work in reference frame K.

The retarded four-force with which particle CD acts on particle AB is

$$\mathbf{f}_{ret} \times AB = \frac{-1}{s_o}\left(\frac{Qq}{2R'^2}\frac{\vec{R}_{ret}}{R_{ret}}AB, i\frac{Qq}{2R'^2}AB\right). \tag{129}$$

The advanced four-force with which particle AB acts on particle CD is

$$\mathbf{g}_{adv} \times CD = \frac{-1}{s_o}\left(-\frac{Qq}{2R_{ret}^2}\frac{\vec{R}_{ret}}{R_{ret}}\frac{CD}{AB}CD, -i\frac{Qq}{2R_{ret}^2}\frac{CD}{AB}CD\right). \tag{130}$$

We notice that $\mathbf{g}_{adv} \times CD = -\mathbf{f}_{ret} \times AB$, since $\frac{1}{R'^2}AB = \frac{1}{R_{ret}^2}\frac{CD}{AB}CD$.

The advanced four-force with which particle EF acts on particle AB is

$$\mathbf{f}_{adv} \times AB = \frac{-1}{s_o}\left(\frac{Qq}{2R''^2}\frac{\vec{R}_{adv}}{R_{adv}}AB, -i\frac{Qq}{2R''^2}AB\right). \tag{131}$$

The retarded four-force with which particle AB acts on particle EF is

$$\mathbf{g}_{ret} \times EF = \frac{-1}{s_o}\left(-\frac{Qq}{2R_{adv}^2}\frac{\vec{R}_{adv}}{R_{adv}}\frac{EF}{AB}EF, i\frac{Qq}{2R_{adv}^2}\frac{EF}{AB}EF\right). \tag{132}$$

We notice that $\mathbf{g}_{ret} \times EF = -\mathbf{f}_{adv} \times AB$, since $\frac{1}{R''^2}AB = \frac{1}{R_{adv}^2}\frac{EF}{AB}EF$.

When going from the worldline string model back to the material point particle model, the product of the linear four-force density with the length of the infinitesimal worldline segment is replaced by the product of the four-force with the infinitesimal change in proper time, an expression equal to the change in four-momentum. The principle of action and reaction that we have in the worldline string model is equivalent to the law of conservation of total four-momentum that we have in the material point particle model.

6   Concluding remarks

This is the third manuscript in which we have investigated a time symmetric action-at-a-distance theory of electrodynamic interaction. Some of the assumptions put forward in the first manuscript [3] have been proven wrong, and they have been replaced with different assumptions in the second manuscript [17]. In the final formulation our theory is based on two postulates.

Postulate I. The interaction is time symmetric, with the retarded and the advanced parts on equal footing. The retarded and the advanced four-forces (the exchanges in four-momentum) are parallel to the displacement four-vectors connecting the two interacting particles.

This means that we implement "the parallel condition".

Postulate II. The interaction takes place between worldline segments of infinitesimal length, whose end points are connected by light signals. In the inertial reference frame in which the source particle is instantaneously at rest, the linear four-force density acting on the test particle is proportional to the product of the Coulombian electrostatic force with the ratio of the length of the infinitesimal segment on the worldline of the test particle to the length of the corresponding infinitesimal segment on the worldline of the source particle.

This is how we take into consideration "the thickness of the light-cone".

In the present work some new geometrical and algebraic derivations are presented, with the goal of making the overall exposition clearer. The action and reaction principle, proven to work in a particular situation in [17], is here demonstrated in the general case. We have also derived the Lorentz invariant expression of our electrodynamic four-force. This proposed four-force (125) matches with a very good approximation, or even exactly (when the relative velocity has the radial direction), the classical expression of the electrodynamic four-force.

We also notice that, in our theory, when the source particle is in uniform or in hyperbolic motion, the four-force acting on the test particle is orthogonal to the four-velocity of the test particle. This orthogonality does not necessarily happen for a source charge in random motion. Thus, in our theory, the rest mass of a particle will change a little bit during interactions. The variation of the rest mass is something that also shows up in special conformal transformations. For a very simple example we can prove that the rest mass is the same before and after the interaction [3]. We hope that this result holds in general.

The remarkable fact about our electrodynamic four-force (125) is that it does not depend on the acceleration of the source particle. However, the classical expression of the four-force [19] shows an explicit dependence on the retarded four-acceleration of the source particle. How can we reconcile these results? We have to remember that the classical calculation is based on the retarded Liénard-Wiechert electromagnetic four-potential. For a source particle at rest, or in uniform motion, or in hyperbolic motion, the advanced four-potential is equal to the retarded four-potential. By assuming that the higher derivatives of the retarded position four-vector of the source particle (the derivative of the four-acceleration, etc.) make no contribution to the electrodynamic four-force, we are able to replace the advanced part of the electrodynamic four-potential with the corresponding retarded part. This is equivalent to the writing of the advanced position four-vector and of the advanced four-velocity of the source particle in the electrodynamic four-force as truncated Taylor series expansions around the retarded position four-vector of the source particle. What would we have to do if the higher derivatives of the retarded position four-vector of the source particle could not be ignored? In such a situation we would have to keep more terms in the Taylor series expansions of the advanced position four-vector and of the advanced four-velocity of the source particle around the retarded position four-vector of the source particle, thus bringing into the expression of the electrodynamic four-force not only the retarded four-acceleration of the source particle, but also the higher derivatives. These extra terms, however, have negligible contributions in all practical experimental situations investigated in classical electrodynamics.

Appendix A. Ratio of corresponding segments

Consider two points connected by a light signal, point C on the worldline of particle 1, with position four-vector $\mathbf{X}_1$, and point A on the worldline of particle 2, with position four-vector $\mathbf{X}_2$. Consider two other points, also connected by a light signal, and infinitely close to the first two points, point D on the worldline of particle 1, with position four-vector $\mathbf{X}_1 + d\mathbf{X}_1$, and point B on the worldline of particle 2, with position four-vector $\mathbf{X}_2 + d\mathbf{X}_2$. We know that

$$\overrightarrow{AC} \cdot \overrightarrow{AC} = (\mathbf{X}_1 - \mathbf{X}_2) \cdot (\mathbf{X}_1 - \mathbf{X}_2) = 0, \tag{133}$$

and that

$$\overrightarrow{BD} \cdot \overrightarrow{BD} = (\mathbf{X}_1 + d\mathbf{X}_1 - \mathbf{X}_2 - d\mathbf{X}_2) \cdot (\mathbf{X}_1 + d\mathbf{X}_1 - \mathbf{X}_2 - d\mathbf{X}_2) = 0. \tag{134}$$

By rearranging terms in the last equation, we have

$$\big((\mathbf{X}_1 - \mathbf{X}_2) + (d\mathbf{X}_1 - d\mathbf{X}_2)\big) \cdot \big((\mathbf{X}_1 - \mathbf{X}_2) + (d\mathbf{X}_1 - d\mathbf{X}_2)\big) = 0. \tag{135}$$

By expanding the scalar product, we obtain

$$(\mathbf{X}_1 - \mathbf{X}_2) \cdot (\mathbf{X}_1 - \mathbf{X}_2) + 2(\mathbf{X}_1 - \mathbf{X}_2) \cdot (d\mathbf{X}_1 - d\mathbf{X}_2) + (d\mathbf{X}_1 - d\mathbf{X}_2) \cdot (d\mathbf{X}_1 - d\mathbf{X}_2) = 0. \tag{136}$$

Due to equation (133), to first order in the infinitesimals we have

$$(\mathbf{X}_1 - \mathbf{X}_2) \cdot (d\mathbf{X}_1 - d\mathbf{X}_2) = 0, \tag{137}$$

an equation that we can also write as

$$(\mathbf{X}_1 - \mathbf{X}_2) \cdot d\mathbf{X}_1 = (\mathbf{X}_1 - \mathbf{X}_2) \cdot d\mathbf{X}_2. \tag{138}$$

With the substitutions $d\mathbf{X}_1 = \mathbf{V}_1 \, d\tau_1$ and $d\mathbf{X}_2 = \mathbf{V}_2 \, d\tau_2$, where $\mathbf{V}_1$ is the four-velocity of particle 1 at point C and $\mathbf{V}_2$ is the four-velocity of particle 2 at point A, the ratio of the two corresponding infinitesimal segments, connected by light signals at both ends, becomes [20]

$$\frac{AB}{CD} = \frac{i \, c \, d\tau_2}{i \, c \, d\tau_1} = \frac{(\mathbf{X}_1 - \mathbf{X}_2) \cdot \mathbf{V}_1}{(\mathbf{X}_1 - \mathbf{X}_2) \cdot \mathbf{V}_2} = \frac{(\mathbf{X}_2 - \mathbf{X}_1) \cdot \mathbf{V}_1}{(\mathbf{X}_2 - \mathbf{X}_1) \cdot \mathbf{V}_2}. \tag{139}$$

We notice that this formula applies regardless of the temporal ordering of the two corresponding segments.

Appendix B. Ratio of Coulombian radii

We consider the two particles from Appendix A, and we assume that segment CD is in the past of segment AB. Let ict be the time axis of an inertial reference frame co-moving with the source particle at C, and let ict' be the time axis of an inertial reference frame co-moving with the test particle at A. From point A we draw a line perpendicular to the ict axis, which intersects it at point T. The length of segment TA is R, and the length of segment CT is iR. From point C we draw a line perpendicular to the ict' axis, which intersects it at point

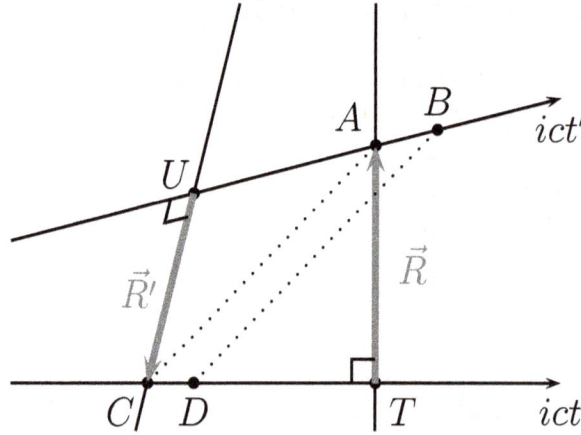

Figure 10: Even for corresponding segments AB and CD that are not in the same Minkowski plane, the relationship $AB/CD = R/R'$ still holds.

U. The length of segment UC is R', and the length of segment UA is iR'. This situation is represented graphically in Fig. 10.

Let K be an inertial reference frame in which particle CD is instantaneously at rest at the origin, defined up to a rotation in its 3D Euclidean space. In this reference frame

$$\mathbf{X}_2 - \mathbf{X}_1 = \overrightarrow{CA} = \overrightarrow{CT} + \overrightarrow{TA} = (\vec{0}, iR) + (\vec{R}, 0) = (\vec{R}, iR), \tag{140}$$

$$\mathbf{V}_1 = (\vec{0}, ic), \tag{141}$$

$$\mathbf{V}_2 = (\gamma_2\,\vec{v_2}, i\gamma_2\,c), \tag{142}$$

$$\tag{143}$$

where $\gamma_2 = \gamma(v_2) = 1/\sqrt{1 - v_2^2/c^2}$. We calculate the scalar products

$$(\mathbf{X}_2 - \mathbf{X}_1) \cdot \mathbf{V}_1 = (\vec{R}, iR) \cdot (\vec{0}, ic) = -Rc, \tag{144}$$

$$(\mathbf{X}_2 - \mathbf{X}_1) \cdot \mathbf{V}_2 = (\vec{R}, iR) \cdot (\gamma_2\,\vec{v_2}, i\gamma_2\,c) = \gamma_2\,\vec{v_2} \cdot \vec{R} - \gamma_2\,Rc, \tag{145}$$

$$\mathbf{V}_1 \cdot \mathbf{V}_2 = (\vec{0}, ic) \cdot (\gamma_2\,\vec{v_2}, i\gamma_2\,c) = -\gamma_2\,c^2. \tag{146}$$

Let K' be an inertial reference frame in which particle AB is instantaneously at rest at the origin, defined up to a rotation in its 3D Euclidean space. In this reference frame

$$\mathbf{X}_2 - \mathbf{X}_1 = \overrightarrow{CA} = \overrightarrow{CU} + \overrightarrow{UA} = (-\vec{R}', 0) + (\vec{0}, iR') = (-\vec{R}', iR'), \tag{147}$$

$$\mathbf{V}_1 = (\gamma_1'\,\vec{v_1'}, i\gamma_1'\,c), \tag{148}$$

$$\mathbf{V}_2 = (\vec{0}, ic), \tag{149}$$

$$\tag{150}$$

46

where $\gamma_1' = \gamma(v_1') = 1/\sqrt{1 - v_1'^2/c^2}$. We calculate the scalar products

$$(\mathbf{X}_2 - \mathbf{X}_1) \cdot \mathbf{V}_1 = (-\vec{R}', iR') \cdot (\gamma_1' \vec{v_1'}, i\gamma_1' c) = -\gamma_1' \vec{v_1'} \cdot \vec{R}' - \gamma_1' R'c, \tag{151}$$

$$(\mathbf{X}_2 - \mathbf{X}_1) \cdot \mathbf{V}_2 = (-\vec{R}', iR') \cdot (\vec{0}, ic) = -R'c, \tag{152}$$

$$\mathbf{V}_1 \cdot \mathbf{V}_2 = (\gamma_1' \vec{v_1'}, i\gamma_1' c) \cdot (\vec{0}, ic) = -\gamma_1' c^2. \tag{153}$$

Since the scalar products are Lorentz invariant, it follows that, in any reference frame, we have

$$\frac{R}{R'} = \frac{(\mathbf{X}_2 - \mathbf{X}_1) \cdot \mathbf{V}_1}{(\mathbf{X}_2 - \mathbf{X}_1) \cdot \mathbf{V}_2}. \tag{154}$$

Together with equation (139), we conclude that

$$\frac{AB}{CD} = \frac{R}{R'}. \tag{155}$$

When segment CD is in the future of segment AB, the segment is renamed EF. In this case $\mathbf{X}_2 - \mathbf{X}_1 = (\vec{R}, -iR)$ in reference frame K, $\mathbf{X}_2 - \mathbf{X}_1 = (-\vec{R}', -iR')$ in reference frame K', and the same formula (154) for the ratio of Coulombian radii is derived. Formula (154) is left invariant by a time reversal operation ($\mathbf{V}_1 \to -\mathbf{V}_1$ and $\mathbf{V}_2 \to -\mathbf{V}_2$), or by a permutation of the two particles ($1 \leftrightarrow 2$ and $R \leftrightarrow R'$).

Appendix C. Formulas for the radial velocities

Since the scalar products from Appendix B are Lorentz invariant, we have

$$-Rc = -\gamma_1' \vec{v_1'} \cdot \vec{R}' - \gamma_1' R'c, \tag{156}$$

$$\gamma_2 \vec{v_2} \cdot \vec{R} - \gamma_2 Rc = -R'c, \tag{157}$$

$$\gamma_2 = \gamma_1'. \tag{158}$$

From equation (158) we see that $v_2^2 = v_1'^2 \equiv v^2$. In the particular case when $\vec{v_1'} = -\vec{v_2}$, the two inertial reference frames K and K' are linked by a Lorentz boost and its inverse transformation.

From equation (157) we get the radial component of the velocity of particle 2, as seen in reference frame K.

$$v_{2rad} = \frac{\vec{v_2} \cdot \vec{R}}{R} = \left(1 - \frac{R'}{\gamma_2 R}\right)c. \tag{159}$$

From equation (156) we get the radial component of the velocity of particle 1, as seen in reference frame K'.

$$v_{1rad}' = \frac{\vec{v_1'} \cdot \vec{R}'}{R'} = \left(-1 + \frac{R}{\gamma_1' R'}\right)c. \tag{160}$$

From equation (159) we write

$$\frac{R'}{R} = \left(1 - \frac{v_{2rad}}{c}\right)\gamma, \tag{161}$$

47

and from equation (160) we write

$$\frac{R}{R'} = \left(1 + \frac{v'_{1rad}}{c}\right)\gamma, \tag{162}$$

where $\gamma = 1/\sqrt{1 - v^2/c^2}$ is the common value seen in equation (158). Due to equation (155), we see that equations (161)-(162) are consistent with equations (11)-(12). By multiplying the above two equations we obtain

$$\left(1 - \frac{v_{2rad}}{c}\right)\left(1 + \frac{v'_{1rad}}{c}\right) = \frac{1}{\gamma^2} = 1 - \frac{v^2}{c^2}. \tag{163}$$

Is it possible for the two radial velocities v_{2rad} and v'_{1rad} to be equal? If $v_{2rad} = v'_{1rad} \equiv v_{rad}$, then from equation (163) it follows that $v_{rad}^2 = v^2$, which means that each particle, relative to the other particle, must move only in the radial direction.

References

[1] Hermann Minkowski, *Spacetime, Minkowski's Papers on Spacetime Physics*, (Minkowski Institute Press, Montreal, 2020), 59 and 164.

[2] Călin Galeriu, "Addition of velocities and electromagnetic interaction: geometrical derivations using 3D Minkowski diagrams", *Apeiron* **10**, 1 (2003).

[3] Călin Galeriu, "Time-Symmetric Action-at-a-Distance Electrodynamics and the Structure of Space-Time", *Physics Essays* **13**, 597 (2000).

[4] Gen Yoneda, "Action and reaction in special relativity", *Eur. J. Phys* **15** (3), 126 (1994).

[5] A. D. Fokker, *Time and Space, Weight and Inertia* (Pergamon, Oxford, 1965), 101.

[6] J. L. Synge, *Relativity: The Special Theory*, (North-Holland, Amsterdam, 1956), 211 and 254.

[7] J. L. Synge, "Angular Momentum, Mass-Center and the Inverse Square Law in Special Relativity", *Phys. Rev.* **47**, 760 (1935).

[8] Vesselin Petkov, *Relativity and the nature of spacetime*, (Springer-Verlag, Berlin, 2005), 58, 77, and 122.

[9] Kevin Brown, *Physics in Space and Time*, (Lulu Press, Research Triangle, 2015), 169 and 228.

[10] A. D. Fokker, "Ein invarianter Variationssatz für die Bewegung mehrerer elektrischer Massenteilchen.", *Z. Phys.* **58**, 386 (1929).

[11] John A. Wheeler and Richard P. Feynman, "Classical Electrodynamics in Terms of Direct Interparticle Action", *Rev. Mod. Phys.* **21**, 425 (1949).

[12] M. Amaku, F. A. B. Coutinho, O. Éboli, and E. Massad, "Some problems with the use of the Dirac delta function I: What is the value of $\int_0^\infty \delta(x)\,dx$?", *Rev. Bras. Ens. Fis.* **43**, e20210132 (2021).

[13] Călin Galeriu, "The algebraic origin of the Doppler factor in the Liénard-Wiechert potentials", *Eur. J. Phys.* **44**, 035203 (2023).

[14] H. Tetrode, "Über den Wirkungszusammenhang der Welt. Eine Erweiterung der klassischen Dynamik.", *Z. Phys.* **10**, 317 (1922).

[15] Nicholas Wheeler, Electrodynamics lecture notes, chapter 6.
`http://www.reed.edu/physics/faculty/wheeler/documents/`
`Electrodynamics/Class%20Notes/Chapter%206.pdf`

[16] Olivier Costa de Beauregard, "Dynamique relativiste des n points et statique classique des n fils", *Comptes rendus de lI Académie des Sciences* **237**, 1395 (1953).

[17] Călin Galeriu, "Electric charge in hyperbolic motion: arcane geometrical aspects", arXiv:1712.02213 [physics.gen-ph] (2017).

[18] Călin Galeriu, "An Introduction to Minkowski Space", *Mathematical Spectrum* **36**, 5 (2003).

[19] Călin Galeriu, "Electric charge in hyperbolic motion: the early history", *Arch. Hist. Exact Sci.* **71**, 363 (2017).

[20] Colin LaMont, "Relativistic Direct Interaction Electrodynamics: Theory and Computation", (B.A. Thesis, Reed College, Oregon, 2011), 39.

3 Analogue Gravity and the de Broglie wave: a Missed Opportunity

Daniel Shanahan

Abstract In a small book entitled *Ondes et Mouvements* [1], published in February 1926, Louis de Broglie described the wave, now known as the de Broglie wave, as a modulation or beating effect of undulatory form induced in the structure of the particle by the failure of simultaneity. In this interpretation, the de Broglie wave is neither ontologically distinct, nor in any way separate, from the particle, but like the Fitzgerald-Lorentz contraction is a distortion in the structure of the particle itself. So understood, the de Broglie wave is a physically real phenomenon, describing for the particle, a well-defined and physically realistic trajectory, whereas the wave functions that emerge as solutions to the Schrödinger and Klein-Gordon equations would be better regarded as mathematical constructs, albeit constructs of significant utility, identifying the wave number and frequency that the particle *would* have at each point of space if it *were* in fact at that point of space. I show that the de Broglie wave would emerge as such a distortion in certain sonic quasiparticles proposed in the context of analogue gravity for the purpose of simulating the Lorentz transformation. I also show how this understanding of the wave would explain, not only its physical nature, but the consistency of its superluminal velocity with special relativity, why the wave emerges only when the particle is observed to be moving and why the particle scatters in accordance with its de Broglie wavelength rather than its Compton wavelength. I discuss the importance of this understanding of the wave for the interpretation of quantum mechanics and the pursuit of quantum gravity.

Keywords Analogue gravity · quantum gravity · Minkowski spacetime · Lorentz transformation · de Broglie wave · wave function

1 Introduction

This presentation concerns an opportunity - a missed opportunity - to employ the methods of analogue gravity to demonstrate the physical origin of the de Broglie wave as that origin was described by de Broglie himself in a small book entitled *Ondes et Mouvements* [1] published in February, 1926.

In *Ondes et Mouvements*, which he completed little more than a year after his better known thesis [2], de Broglie described the wave, not as a true wave, but as a modulation or beating effect of undulatory or sinusoidal form, induced in the underlying structure of the particle by the failure of simultaneity. Understood in this way, the de Broglie wave is not something ontologically distinct, or in any way physically separate, from the particle but, like the Fitzgerald-Lorentz contraction, is a distortion in the structure of the particle itself.

Kyley Ewing (Ed.), *Spacetime Conference 2024. Selected peer-reviewed papers presented at the Seventh International Conference on the Nature and Ontology of Spacetime, 16 - 19 September 2024, Albena, Bulgaria* (Minkowski Institute Press, Montreal 2025). ISBN 978-1-998902-44-6 (softcover), ISBN 978-1-998902-45-3 (ebook).

I will discuss two analogues that employ the methods of analogue gravity to simulate, not the Hawking radiation that has been the primary interest of analogue gravity, but the effects of the Lorentz transformation. In these *sonic* analogues, the role of the speed of light is played instead by that of sound. One such analogue is described in a paper by Barceló and Jannes [3], and the other is due to Todd and Menicucci [4]. These analogues show how changes predicted by the Lorentz transformation might be simulated in a universe in which everything within the universe, including particles and forces and observers and measuring devices, is formed from sonic waves[1].

These curious universes have been referred to as fishbowl universes, it being possible to contemplate two kinds of observer, one within the fishbowl where everything is constructed from effects that evolve at the speed of sound, and the other who is outside the fishbowl and like some supernatural being is able to look into the bowl to observe the strange workings of (sonic) special relativity. Of course this god-like observer may simply be the post-doc who is running the experiment, but from their privileged position outside the bowl, an *external* observer will be able to perceive how changes of length, time and simultaneity ensure that the speed of light and the laws of physics are the same for all *internal* observers.

Neither of the two papers actually mentions the de Broglie wave. Yet the de Broglie wave is also a consequence of the Lorentz transformation, specifically of the failure of simultaneity. I will show that if the sonic analogue of a massive particle were constructed in the manner described in either of these sonic analogues, it would engender the de Broglie wave in precisely the manner described in *Ondes et Mouvements*.

It is important to notice that this way of understanding the de Broglie wave is not an alternative to some orthodox or "standard" explanation of the wave. Standard quantum mechanics (SQM) has no such explanation. If you were to consult a standard text or ChatGPT or Wikipedia, you would be referred to the concept of wave-particle duality and would learn that a particle acts sometimes as a particle and sometimes as a wave, and that the latter serves, in some mysterious manner, as a wave of probability.

You would also be told (quite correctly) that this wave or wave-like phenomenon has a frequency ω_E directly related to the energy E of the particle by the Planck-Einstein relation,

$$E = \hbar\omega_E = \hbar\gamma\omega_o, \tag{1}$$

and a wave number κ_{dB} similarly related to the momentum p of the particle by the de Broglie relation,

$$p = \hbar\kappa_{dB} = \hbar\gamma\,\omega_o\frac{v}{c^2}, \tag{2}$$

where $\hbar$ is the reduced Planck's constant, ω_o is the natural or characteristic frequency of the particle at rest, v is the velocity of the particle, and c is the speed of light in vacuum.

All this is standard fare, but these are not explanations. You would not learn what this wave actually is, nor why a particle sometimes behaves as a wave and sometimes as a

---

[1]For discussions of the implications of these sonic analogues for the fundamentality of the speed of light, see Cheng and Read [5] and Shanahan [6].

particle, nor why the wave is superluminal, nor why it emerges only when the particle is moving. The answers to all these questions *are* apparent in *Ondes et Mouvements*.

Since 1926, this explanation of the de Broglie wave has been discovered and rediscovered many times, and now has a modest literature[2]. Thus a question I will need to address is why, if the interpretaion presented in this literature is physically reasonable and has no apparent alternative, it has not yet achieved the status of orthodoxy.

But before proceeding further, I should explain what I mean by a modulation or beat induced by the failure of simultaneity.

2 Beats, simultaneity and the de Broglie wave

A beat or phase modulation[3] is a periodic variation in intensity caused by interference between two waves of different frequency. Its occurrence in music was studied by the ancient Greeks (see Lindsay [8]), while its origin in interference, in both sound (Rayleigh [9]) and light (Brewster [10]), has been well-understood mathematically since at least the 19th century.

An illustration of how such a modulation is induced by a failure of simultaneity is provided by the standing wave of Fig. 1. In its rest frame, every part of the standing wave is oscillating in unison as in Fig. 1(a). But to an observer for whom the frame of the wave is moving to the right at a relativistic velocity, as in Fig. 1(b), the standing wave is experiencing the changes described by the Lorentz transformation. These include the failure of simultaneity. To the stationary observer, those parts of the wave to the right are rising and falling later than those to the left.

If the inertial frame of the wave is moving even faster relative to the observer, as in Fig. 1(c), this progressive retardation in phase will be observed as a sinusoidal wave advancing through the underlying wave structure, and having the velocity and other characteristics of the de Broglie wave.

As will be discussed in the next section, the de Broglie wave described by de Broglie in *Ondes et Mouvements* [1] is the realization in three dimensions of the progressive retardation of phase depicted in Fig. 1(c). And as will then be discussed in Sect. 4, the quasiparticles of the two sonic analogues simulate the Lorentz transformation because they too are structured as standing waves, from which the de Broglie wave would thus emerge as a modulation as the particle moves.

The simple standing wave of Fig. 1 already suffices to explain a number of features of the de Broglie wave that would be anomalous in a true wave. One is the velocity of the modulation, which becomes infinite as the particle comes to rest because all parts of the

---

[2]See the listing attempted in Ref. [7], as well as two more recent papers of my own, published last year, one in *Foundations of Physics* [6], the other pursuant to a presentation at a conference at the Sorbonne commemorating the centenary of de Broglie's first papers on the wave [11].

[3]I have not burdened this presentation with the mathematical analysis of standing waves and beats, other that that due to de Broglie himself, some of which is presented in a simplified form in Sect. 3.

wave are then oscillating in unison. The progression of phase from one peak to the next is now instantaneous, and the modulation disappears.

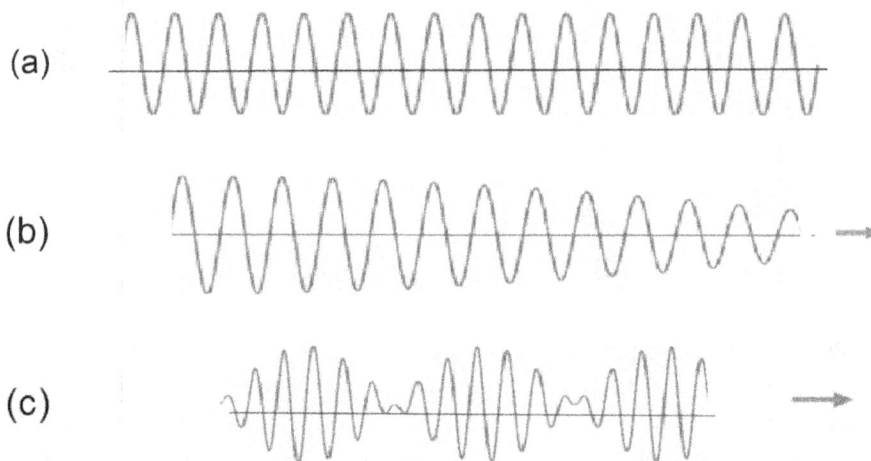

Figure 1: (a) a standing wave (b) the same wave moving to the right and progressively retarded in phase in that direction as a consequence of the failure of simultaneity (c) the same wave now moving at a sufficiently greater velocity that the progressive retardation of phase is observed as a sinusoidal beat or phase modulation having the velocity and other characteristics of a de Broglie wave.

It should also be apparent from the lowermost drawing why the manner in which a massive particle interferes is determined by the wavelength of the de Broglie wave. It will be in its sinusoidally distorted form that the moving particle approaches and interacts with a device such as a (stationary) beam splitter. Whether its interference at a beam splitter with another such wave is constructive or destructive, or somewhere in between, will thus depend on the degree to which the modulations are in or out of phase at the beam splitter.

There are two reasons why interference is not also observed at the Compton wavelength, one being that the Compton wavelength is much smaller than the separations of typical scattering elements, the other being that the Compton wavefronts are distributed in a spheroidal manner about the centre of the particle and unlikely to have any coherent relationship with the spacings of those scattering elements.

These drawings also reveal the relevance of the de Broglie wave to the covariance ensured by the Lorentz transformation. It is the full modulated wave structure, rather than the de Broglie wave considered *solus,* that is the covariant relativistic object. The Fitzgerald-Lorentz contraction appears in the carrier wave (3), while the dilation of time and failure of simultaneity predicted by the Lorentz transformation are described by the modulation, that is to say, by the de Broglie wave (5).

Above all, these drawings illustrate in one dimension the ontological parsimony of this understanding of the de Broglie wave. A three-dimensional structure due to de Broglie himself will be described in the next section. In neither case, does the wave require further structure or new physics. There is no need to rationalize the superluminality of the wave, or to consider the ontological status of a wave that might otherwise seem to emerge from

54

nowhere. The de Broglie "wave", so called, is simply, as stated above, a distortion predicted in well known manner by the Lorentz transformation, and entirely consistent with classical wave theory.

3 *Ondes et Mouvements*

In missing the de Broglie wave, the originators of the two sound analogues were in illustrious company. Its existence was missed by Einstein in 1905 [12] and Minkowski in 1908 [13] and again by Einstein when he presented his general theory in 1916 [14].

When in 1923 de Broglie did propose this wave, it proved elusive in yet another sense. He first described the wave - in a short paper [15] published prior to his thesis - as fictitious (*un onde fictive*). In a subsequent note [16], it had become a "phase wave", and by the time of the famous thesis [2], this phase wave had acquired physical reality, albeit as a curiously superluminal wave that seemed to be related in some way to a spatially extended "periodic phenomenon", which de Broglie described as surrounding the particle in its rest frame.

But in the thesis, de Broglie derived this phase wave, not by relating it to the periodic phenomenon surrounding the particle, but from his "theorem of the harmony of phases", according to which, the superluminal wave maintains consistency of phase with the subluminal particle as it moves.

That de Broglie was aware of the provisional state of his doctoral submission is evident from its very last paragraph [2]:

> I have left the definitions of phase waves and the periodic phenomena for which such waves are a realization .... deliberately vague. The present theory is, therefore, to be considered rather tentative as physics and not an established doctrine.

But with his doctorate in hand, de Broglie embarked upon *Ondes et Mouvements* [1], in which what had been "deliberately vague" and prudently "tentative" the previous year now took a more decided form. In *Ondes et Mouvements*, the periodic phenomenon is clearly a standing wave, from which the de Broglie wave emerges as what is also very clearly now, not a true wave, but the relativistically induced phase modulation of the underlying wave structure.

De Broglie showed that under a Lorentz boost,

$$x' = \gamma \left(x - vt \right),$$
$$y' = y,$$
$$z' = z,$$
$$t' = \gamma \left(t - \frac{vx}{c^2} \right),$$

a spatially extended "periodic phenomenon" of any form,

$$f(x', y', z') \, e^{i\omega_0 t'},$$

that is oscillating in three dimensions at some frequency ω_0 becomes,

$$f\left(\gamma(x - vt), y, z,\right) e^{i(\omega_E t - \kappa_{dB} x)},$$

in which the spatial factor $f(x', y', z')$ is now the carrier wave,

$$f\left(\gamma(x - vt)\right) \tag{3}$$

which is moving in the x-direction at velocity v, and as can be seen from the inclusion of the Lorentz factor γ, has contracted in that direction in accordance with the Fitzgerald-Lorentz contraction.

What is of particular relevance here is that the oscillatory factor $e^{i\omega_0 t'}$ is now,

$$e^{i\omega_0 \gamma(t - vx/c^2)}, \tag{4}$$

and describes a progressive loss of phase (the modulation of beating effect discussed above) evolving through the carrier wave (3) at the superluminal velocity,

$$v_{dB} = \frac{dx}{dt} = \frac{c^2}{v}.$$

With the assistance of Eqns. (1) and (2), Eqn. (4) can be rewritten in terms of the Einstein frequency ω_E and de Broglie wave number κ_{dB} as,

$$e^{i(\omega_E t - \kappa_{dB} x)}, \tag{5}$$

and is now more readily recognizable as the de Broglie wave.

Combining wave factors (3) and (5), the full modulated wave is,

$$f\left(\gamma(x - vt), y, z\right) e^{i(\omega_E t - \kappa_{dB} x)}. \tag{6}$$

But it is not every conceivable massive structure to which the Lorentz transformation can be *validly* applied. One example of an object to which the transformation cannot be applied and which accordingly could not exist in our universe is a perfectly rigid object. Such a structure has spatial extension and could be oscillatory, but the displacement of one end would be transmitted instantaneously to the other, which would be contrary to the limiting role of the speed of light. The question of what can and cannot have physical existence in the actual universe has a direct bearing on how an object might be plausibly analogized, and this a question to which I will return in Sect. 6.

Before leaving the present section, I mention a more specific model discussed by de Broglie in *Ondes et Mouvements* in which the "periodic phenomenon" has the idealized form of a spherical standing wave,

$$\varphi(\mathbf{r'}, t') = \frac{A}{|\mathbf{r'}|} \sin(\kappa_0 r') e^{i\omega t'}, \tag{7}$$

where,

$$\mathbf{r'} = \sqrt{x'^2 + y'^2 + z'^2}.$$

Under a Lorentz transformation, and switching now to cartesian coordinates, this wave structure becomes[4],

$$\varphi(x,y,z,t) = \frac{A}{\sqrt{\gamma^2(x-vt)^2 + y^2 + z^2}} \sin\!\left(\kappa_0\sqrt{\gamma^2(x-vt)^2 + y^2 + z^2}\,\right) e^{i(\omega_E t - \kappa_{dB}x)}, \quad (8)$$

where the previously spherical structure has contracted in the direction of motion and acquired the form of an oblate spheroidal wave.

The idealized model described by Eqn. (7) lacks the asymmetries that in an actual particle are suggested by properties such as charge, spin and parity, but displays in a conveniently simplified form, properties that I will argue in Sect. 6 are essential to the structure of all massive particles, namely that the particle be not only spatially extended and oscillatory, but that it comprise underlying influences and effects evolving at velocity c.

This model particle is also consistent with the assertion in Sect. 1 of this paper that the de Broglie wave is not in any way separate or ontologically distinct from the particle, but a distortion induced by the failure of simultaneity in the structure of the particle itself.

4   The fishbowl universes

Interest in sonic analogues as a means of investigating Hawking radiation ([17] [18]) seems to have been sparked by Unruh's suggestion in 1981 that the behaviour of a quantum field in a classical gravitational field might be modelled by sound waves in a convergent fluid flow (Unruh [19]).

The idea is to create what has been referred to as a "dumb hole", which in the parlance of analogue gravity is a region from which sound cannot escape, and which is formed by having the medium carrying that sound move at a speed greater than that of the sound itself. Unruh showed that near its event horizon, the metric of such a model would correspond with the Schwarzschild metric. He suggested that while the possibility of actually constructing such an analogue might be "extremely slim", it would present a simpler task than that of creating an actual black hole or of finding a small black hole near the Earth!

Analogues employing various species of sonic waves in various media have since been proposed (see the review by Barceló et al [20]) and an impressive number have actually been constructed, beginning with that of Steinhauer [21] in 2016, who reported the observation of Hawking radiation in the sonic analogue of a one-dimensional black hole formed in a Bose-Einstein concentrate.

Analogue gravity now has a significant literature. Reviews have been written and books and anthologies have been published. Arguments both for and against the confirmatory

---

[4]I have simplified slightly de Broglie's equations and have expressed the final wave factor in Eqn. (8) in the more usual form of the de Broglie wave.

value of these analogues can be found in this literature, of which two interesting examples are Dardashti et al [22] (for) and Crowther et al [23] (against).

However, my concern here is not with black holes and Hawking radiation, but with a small subset of this literature relating to what might be termed analogue special relativity or analogue Minkowski spacetime, and in particular, as mentioned above, the sonic analogues proposed by Barceló and Jannes [3] and Todd and Menicucci [4].

Imagine a universe, or simply a closed laboratory (a fishbowl universe), in which the velocity c_s of sound is measured by devices formed from material in which all physical influences evolve at the velocity of sound. In the analogue of Barceló and Jannes, the arms of a Michelson-Morley interferometer comprise equally spaced quasiparticles, which in a collective oscillation produce waves that evolve outwardly in all directions at the velocity c_s. When this sonic interferometer moves at a velocity $v < c_s$ with respect to the surrounding medium, it experiences a relativistic contraction with a sonic Lorentz factor,

$$\gamma_s = (1 - \frac{v^2}{c_s^2})^{-\frac{1}{2}}, \qquad (9)$$

based on the speed c_s of sound rather than the speed c of light.

Todd and Menicucci [4] show how *all* three of the curious changes predicted by the Lorentz transformation, namely the contraction, the dilation of time and the failure of simultaneity, might be analogized by a chain of sound clocks, these being akin to the light clocks described by Einstein except that the return journey between opposed mirrors is made by sound waves rather than electromagnetic waves.

These curious fishbowl universes differ considerably of course from the universe we know, and the significance of these disanalogies will be discussed in Sect. 6. But for the purposes of the present section, all that need be noticed is the central assumption of these analogies, which is that everything within the actual universe, including observers and their measuring instruments, can be plausibly analogized by structures comprising counter-propagating sound waves of velocity c_s. As the quasiparticles of Barceló and Jannes propagate outwardly, each will be the recipient of incoming waves from other quasiparticles, while in the chains of sound clocks described by Todd and Menicucci, these counter-propagating waves comprise sequences of sound pulses making return trips between opposed reflectors.

The massive quasiparticles of these fishbowl universes thus comprise structures that, when at rest in the medium, are akin to standing waves, or sufficiently so that they transform between one inertial frame and another in the manner described by de Broglie in *Ondes et Mouvements* [1]. It follows therefore that while the de Broglie wave may not have been within the contemplation of the originators of these analogues, this wave-like phenomenon should emerge in the quasiparticles of each analogue if it were actually built and tested, a possibility that I will also say something about in Sect. 6.

Why then was the de Broglie wave not noticed in the formulation of these analogues? One reason, of course, is that the objective was merely to analogize the Lorentz transformation, and in that objective these sonic analogues have succeeded very well. But it is also relevant to notice here that these analogues were effectively pre-quantum and would have inevitably missed the de Broglie wave for very much the same reason that it was missed by Einstein in 1905 [12]. In each analogue, it was important for the authors to show that their measuring devices, that is to say, the sonic interferometer of Barceló and Jannes, and the chain of sound clocks of Todd and Menicucci, were plausible simulations of the corresponding devices of the actual universe. In deriving the Lorentz transformation, they were thus concerned, not with the microscopic structure of matter, but with the plausibility of devices capable of simulating the macroscopic measuring rods and clocks considered by Einstein in 1905.

The derivations thus proceeded, as Einstein did in 1905, from a consideration of the classical rather than the quantum. Todd and Menicucci [4] inform the reader that their sound clocks are separated by "spacing arms", while Barceló and Jannes [3], at 194, employ "emergent vector fields and sources to produce a rigid bar". In each case, the derivation involved the consideration, not of changes in the frequencies and wave numbers of counterpropagating waves, but of the differing times taken by light, propagating longitudinally and transversely, with respect to the direction of motion of the macroscopic measuring device in question.

In 1905, Einstein was unaware of the de Broglie wave. While he realized that if light is to have the same velocity c for all observers, solid matter must change in the manner of counterpropagating light rays, he was unable to take the further step of proposing that in some sense solid matter must in fact comprise counterpropagating wave-like influences of velocity c.

When Einstein did learn from de Broglie's thesis of the wave-like behaviour of matter, he famously declared, in a latter to Paul Langevin, that de Broglie had "lifted a corner of the great veil" (as cited in Ref. [24]).

5 Schrödinger's wave functions

Unfortunately, for the orderly development of quantum mechanics, *Ondes et Mouvements* [1] was overtaken in early 1926 by rapidly developing events, these being the publication in quick succession of Schrödinger's papers on wave mechanics (Schrödinger [25]) and Born's proposal that the wave functions from the Schrödinger equation are objectively probabilistic (Born [26]). This was unfortunate, at least, for any prospect of a physically realistic interpretation of quantum mechanics consistent with the suggestion in *Ondes et Mouvements* that the de Broglie wave is a physically real phenomenon that describes for the particle, a well-defined and physically realistic trajectory.

Schrödinger's papers attracted immediate interest. He was able to explain the observed energies of the Hydrogen atom and harmonic oscillator, as also the Stark and Zeeman effects. Crucially, he demonstrated the equivalence of his wave mechanics and the earlier matrix mechanics of Heisenberg, Born and Jordan (Schrödinger [27]).

The first of Schrödinger's papers on wave mechanics (Schrödinger [28]) was published just three weeks prior to *Ondes et Mouvement*s and was apparently written in ignorance of that work. Indeed, there is no reference to *Ondes et Mouvements* or the interpretation of the de Broglie wave as a modulation in any of the papers that Schrödinger submitted to Annalen der Physik during 1926 and 1927. While Schrödinger acknowledged in his papers his intellectual debt to de Broglie, the de Broglie wave seems to have remained for Schrödinger the "intentionally vague" and "tentative" superluminal wave of de Broglie's thesis of 1924 [2].

It was inevitably the wave function rather than the de Broglie wave that now became the focus of enquiry, and it would seem that from this time de Broglie's own efforts were concentrated on reconciling his ideas with those of Born and Schrödinger. His double solution paper of 1927 (de Broglie [29]) is very much concerned with the interpretation of Schrödinger's wave functions, which de Broglie treated as both the source of a guiding function and a means of determining the "probability of presence" of the electron in a manner which, as he put it in that paper, "approaches the one brilliantly upheld by Born" (and see also de Broglie [30]).

Yet despite the predictive value of those wave functions, they have seemed sufficiently mysterious as to encourage a debate as to whether they are ontic or merely epistemic in their significance, that is to say, whether they are physically real waveforms, or merely a means of calculating energies and momenta, for which, as Schrödinger demonstrated, they are highly successful. And of course, it is the objectively probabilistic Born interpretation of these wave functions that has been the source of the measurement problem of quantum mechanics.

Let us assume then that de Broglie may have been correct in the interpretation of the de Broglie wave he presented in *Ondes et Mouvements* and consider what this interpretation might have to say regarding the physical meaning of these curious wave functions.

It would seem that as they were originally conceived, the Schrödinger and Klein-Gordon equations were intended as equations for the de Broglie wave (see Jammer [31], at p. 255 et seq. and Bloch [32]). As the story goes, Schrödinger led a colloquium on de Broglie's thesis, and during the ensuing discussion, one of his audience, apparently the Dutch physical chemist Peter Debye, commented that talk about a wave was a bit silly in the absence of an equation for the wave.

We can imagine that Schrödinger was a little peeved at this, but he did then develop a wave equation - apparently during a romantic sojourn in the Swiss alps (Moore [33], at p. 140 et seq.). He first derived the relativistic equation, which is to say the aforesaid Klein-Gordon equation, since named after two subsequent discoverers. But on encountering difficulties with the relativistic equation, he turned instead to the nonrelativistic equation, that is to say the Schrödinger equation, and it was this that Schrödinger presented in the papers on wave mechanics that he submitted to *Annalen der Physik* during 1926 and 1927 (collected in Ref. [25]).

But it would be incorrect to suppose that the Schrödinger and Klein-Gordon equations are equations for the wave contemplated by de Broglie. In his thesis of 1924 [2], and even more clearly in *Ondes et Mouvements* [1], de Broglie described the de Broglie wave as centred upon the position of the moving particle. As the particle moved, the spatial evolution of its wave would thus define a trajectory for the particle.

However, it is apparent from Schrödinger's papers, as also from the equation itself, that the wave for which Schrödinger constructed the Schrödinger equation was not the localized travelling wave contemplated by de Broglie, but what would be better described as a mathematical construct - an artificial wave having at every point within its domain, the frequency and wave number that *would* be associated with a particle of a specified energy if it *were* at that same point of space.

That this is so can be understood from the manner in which the Schrödinger equation is derived from the corresponding classical equation of motion,

$$E = \frac{p^2}{2m} + V, \tag{10}$$

where, by employing the Planck-Einstein and de Broglie relations (Eqns. (1) and (2), respectively) to make the substitutions,

$$E \rightarrow i\hbar\frac{\partial\psi}{\partial t}, \quad and \quad \mathbf{p} \rightarrow i\hbar\nabla\psi, \tag{11}$$

we obtain,

$$i\hbar\frac{\partial\psi}{\partial t} = -\frac{\hbar^2}{2m}\nabla^2\psi + V\psi = 0, \tag{12}$$

which is the Schrödinger equation (and by making the same substitutions in the corresponding relativistic equation of motion we obtain, in the same manner, the Klein-Gordon equation).

Clearly, the classical equation of motion (Eqn. (10)) does not identify a particular trajectory. It is a general rule governing all possible (non-relativistic) trajectories for a mass of a particular energy in a given potential. It is thus nomological in its primary significance, yet at the same time epistemic, for if the vector momentum of an object is known at some point within a known potential field, its trajectory may then be deduced.

So too, the Schrödinger and Klein-Gordon equations should be regarded as nomological and epistemic. As de Broglie suggested in his double solution paper of 1927 [29], a solution to the Schrödinger equation may be thought of as representing, not a single trajectory, but the trajectories of a "swarm of particles" filling the entire domain of the problem. As de Broglie also showed, an individual trajectory can be deduced from its wave function by simply reversing the second of the substitutions (11) to obtain a guidance equation, which might simply have the form (see for example de Broglie [34], at p.94),

$$\mathbf{p} = i\hbar\frac{\nabla\psi}{\psi}.$$

And that I suggest is very largely why these wave functions have seemed so inscrutable. A wave function has by construction, at each point of space, the frequency and wave number that in accordance with the Planck-Einstein and de Broglie relations, correspond with the energy and momentum, respectively, that the particle would have at that same point of space. But the wave equation is agnostic as to the existence of this particle. Occupying, in the manner of a wave, the entire space available to a wave, a wave function will display a symmetry of structure from which it may not even be apparent that the particle has a trajectory, let alone a well-defined and physical reasonable trajectory of the kind supposed in physically realistic interpretations of quantum mechanics.

The distinction that I am making here between wave function and de Broglie wave can be illustrated in a rather stark manner by imagining a molecule, perhaps of oxygen, that is somewhere in the room, but we don't know where. Because the Schrödinger wave function covers all possible locations, it must encompass in like manner the entire room. But wherever the molecule might actually be, its de Broglie wave will be a microscopic wave centred on the current position of the particle with an amplitude increasingly attenuated with distance from that position.

While de Broglie supposed the existence of both wave and particle, Schrödinger took the position at the time that the physical entity is the wave, that is to say, the wave function. For Schrödinger, there was only the wave! In his address to the Solvay meeting of 1927, he explained this as as follows:

> I myself have so far found useful the following perhaps somewhat naive but quite concrete idea. The classical system of material points does not really exist, instead there exists something that continuously fills the entire space and of which one would obtain a 'snapshot' if one dragged the classical system with the camera shutter open through *all* its configurations, the representative point in q-space spending in each volume element $d\tau$ a time that is proportional to the instantaneous value of $\psi\psi^*$. (see Bacciagaluppi and Valentini [35], at p. 411)

It is possible to discern in these differing interpretations, a progression of ontologies from de Broglie, who supposed a localized wave surrounding a localized particle, to Schrödinger who was willing to forsake the particle in favour of an all-encompassing wave function, and from thence to Born, who was unwilling to discard the particle, but willing to abandon the certainties of physical reality in favour of a wave function that would serve as a probability function for the particle. In this confused melee of competing possibilities, there could be wave and particle, or simply the wave, which could be interpreted deterministically or probabilistically, and if probabilistically, objectively so or subjectively so.

In the aftermath of the Solvay conference of 1927, it was Born's objectively probabilistic interpretation of the wave function that eventually achieved orthodoxy. But one possibility that seems to have been overlooked at the time was that de Broglie had correctly interpreted the nature of the de Broglie wave in *Ondes et Mouvements*, but was incorrect in his insistence that wave and particle are separate entities.

It is that possibility that is suggested by the fishbowl universes of Barceló and Jannes [3] and Todd and Menicucci [4] and it is to these that I now return.

6   Analogy and disanalogy

I suggest that in *Ondes et Mouvements*, de Broglie presented the only interpretation of the de Broglie wave that makes physical sense.   The question I now consider is whether, by facilitating an argument by analogy, these ingenious sonic analogues of Barceló and Jannes [3] and Todd and Menicucci [4] might provide the means of resuscitating that explanation.

The idea would be to simulate the Lorentz transformation of a sonic quasiparticle and look for the scattering of that particle in a manner consistent with its de Broglie wave, this being the way in which the actual wave was originally confirmed by the experiments of Davisson and Germer [36] and Thompson [37].

While such an argument from analogy is rarely conclusive [5], the literature suggests that it may well be persuasive depending on the existence or otherwise of significant disanalogies and the degree to which the *source* (the established phenomenon) and the *target* (the hypothesized phenomenon) correspond, see generally Bartha [41].   Analogical reasoning may be particularly plausible, it would seem, when, as with these fishbowl universes, source and target have a common mathematical structure, as is the case, for example, in those situations, ubiquitous in Nature, where two otherwise dissimilar systems exhibit harmonic motion (an example cited by Crowther et al [23]).

The formulation of an appropriate analogue and *a fortiori*, its experimental realization, would address two obstacles to the interpretation of the de Broglie wave proposed in *Ondes et Mouvements*.   By showing how the de Broglie wave might emerge from a thoroughly wave-structured particle, it would provide a practical demonstration that, contrary to de Broglie's position on this particular issue (see, for instance, de Broglie [42]), wave and particle need not be separate physical entities.   By showing that the wave thus emerging is neither fictitious nor probabilistic, but the result of a well-established and well-understood process of interference, it might encourage the adoption of a physically realistic solution to the measurement problem.

For such an analogue to be plausible, there must be reason to assume that in the actual universe, as in the fishbowls, a massive particle may be treated, for the purposes of the Lorentz transformation, as comprising a superposition of waves of a single fundamental velocity.   I will argue that this is implied by two well-established and fundamental principles of physics, the aforesaid Lorentz transformation and the Planck-Einstein relation (Eqn. (1) above).

The Lorentz transformation implies that whatever the structure of a massive particle might be, the various effects from which that structure is constituted, whether they be internal forces, topologies, or whatever, must evolve at the velocity c.   If there were some

---

[5]Even so, some such arguments seem compelling: Galileo inferred the existence of mountains on the moon from his observation that, as on the Earth, points of light appear ahead of the advancing edge of sunlight [38]; Darwin drew support for the hypothesis of natural selection from the analogy of artificial selection [39]; and Priestley argued from the absence of a electric field within a uniformly charged spherical shell that, as in the analogous case with the gravitational field, electrostatic charge must follow an inverse square law [40]. (For these and other illustrations, see Bartha [41]).

fundamental effect or influence that evolved at a velocity other than c, it would have its own Lorentz factor γ and corresponding Lorentz transformation, and neither the structure of matter, nor the laws of physics, could survive unchanged from one inertial frame to the next.

As discussed earlier, it is implicit in Einstein's various thought experiments involving moving trains and railway platforms and the like that this must indeed be the nature of solid matter (see also Shanahan [6] and [43]). Velocities that differ from c, those for example of sound waves, refracted light and massive objects may then be explained as the net effect of underlying influences that *do* evolve at velocity c.

Meanwhile, the Planck-Einstein relation,

$$E = \hbar\omega,$$

suggests that whatever the standard model might ultimately have to say regarding the structures of the elementary particles, these consistory influences of velocity c must have, in the rest frame of the particle, the characteristic frequency ω_o of the species of particle in question.

There are also disanalogies that should be considered. One is that sound requires a medium, which is to say, an analogue of the *luminiferous aether* and "absolutely stationary space" that Einstein dismissed as superfluous in 1905 [12]. Barceló and Jannes [3] and Todd and Menicucci [4] dismiss this as being a significant disanalogy on the basis that the two relativities, that of Einstein and that of Lorentz, are mathematically and empirically equivalent. In this, I believe they are correct.

Two further disanalogies, both of which were referred to earlier in this paper, are not so much reasons to doubt the analogy, but the reasons it is useful. One is that, as Unruh said in 1981, "all the basic physics [of sound] is completely understood" [19]. The other is that the relatively low velocity of sound makes it possible to consider the workings of the Lorentz transformation from the standpoint of an external observer. It follows that if a sonic de Broglie wave were observed, there would be little doubt as to its origin whereas, at the mysterious level of the quantum, there must be at least a theoretical possibility that the de Broglie wave has an origin as yet unknown.

It is apparent from the literature of analogue gravity that to a significant degree the expertise necessary to construct and test a sonic analogue of the de Broglie wave does already exist. For instance - and at the risk of revealing an experimental naiveté - a simple analogue, in which the carrier wave varies only in one dimension, might simply consist of: (a) opposed plates of the kind used in Steinhauer's experiment [21], which by vibrating at the same frequency at opposite ends of a tubular "fishbowl" would create a standing wave; (b) a means of varying the frequency of vibration of either or both plates so as simulate a moving quasiparticle and accompanying beat, that is to say a de Broglie wave, as contemplated in one spatial dimension in Fig. 1; (c) some way of inserting an appropriate scattering device; and (d) a means of confirming the path then taken by the scattered wave.

The scattering element (c) might comprise, for example, parallel wires or thin rods set in the path of the wave, angled as to simulate the scoring of a diffraction grating, and having spacings of an order of magnitude adapted to the scattering of the particle at its de Broglie wavelength rather than its Compton wavelength.

7 Concluding discussion

On comparing the situation of the de Broglie wave with that of Hawking radiation, one might question whether the interpretation in *Ondes et Mouvements* [1] should need further demonstration. Whereas Hawking radiation is empirically inaccessible, controversial and predicted from a relatively abstruse mathematical analysis, the de Broglie wave is evidenced routinely in interferometry and scattering experiments, while the manner in which a beat emerges as a consequence of interference is well-known and understood, and not at all controversial.

And, as I have stressed above, the issue is not as to which of two rival theories provides the better explanation. For the de Broglie wave, there is only the one physically reasonable explanation, unless at least the imprimatur of orthodoxy is to be accorded to a superluminal wave of unknown origin and ontology that seems to arise out of nowhere as the particle moves. The difficulty here is of a different kind, namely that the explanation presented in *Ondes et Mouvements* would be an embarrassment to a quantum theory that has insisted for nearly a century that, prior to measurement, a particle has no location or trajectory.

Yet it was from the prediction and experimental confirmation of the de Broglie wave that it became possible to formulate this quantum mechanics in which all particles, whether massive or massive, are treated as evolving and interacting in the manner of waves. I suggest that in the absence of an explanation of this wave-like behaviour, no quantum effect can be properly understood including, it should be noticed, the Hawking radiation which was the original motivation for these sonic analogues.

References

[1] L. de Broglie, *Ondes et Mouvements*, Gauthier-Villars, Paris (1926)

[2] L. de Broglie, Doctoral thesis, *Recherches sur la théorie des quanta.* Ann. de Phys. (10) **3**, 22 (1925)

[3] C. Barceló, G. Jannes, A Real Lorentz–Fitzgerald contraction, Found. Phys. **38**, 191 (2008)

[4] S. L. Todd, N. C. Menicucci, Sonic relativity, Found. Phys. **47**, 1267 (2017)

[5] B. Cheng, J. Read, Why Not a Sound Postulate? Found. Phys. **51,** 71 (2021)

[6] D. Shanahan, The Lorentz Transformation in a Fishbowl: A Comment on Cheng and Read's "Why Not a Sound Postulate". Found. Phys. **53**, 55 (2023)

[7] D. Shanahan, Reverse Engineering the de Broglie Wave, International Journal of Quantum Foundations **9**, 44 (2023)

[8] R. B. Lindsay, *The Story of Acoustics*, J. Acoustic Soc. Amer. **39**, 629 (1966)

[9] J. W. Strutt, Lord Rayleigh, *The Theory of Sound*, Cambridge University Press (1877)

[10] D. Brewster, *A Treatise on Optics*, Longman, London (1831)

[11] D. Shanahan, The de Broglie wave as an undulatory distortion induced in the moving particle by the failure of simultaneity, Ann. Fond. Louis de Broglie **48**, 197 (2023)

[12] A. Einstein, Zur elektrodynamik bewegter Korper, Ann. Phys. **17**, 891 (1905)

[13] H. Minkowski, Raum und Zeit, Phys. Zeit. **10**, 104 (1909)

[14] A. Einstein, Die Grundlage der allgemeinen der Relativitätstheorie, Ann. Phys. **354**, 769 (1916)

[15] L. de Broglie, Ondes et Quanta, C. R. Acad. Sci. **177**, 507 (1923)

[16] L. de Broglie, Sur la fréquence propre de l'électron, C. R. Acad. Sci. **180**, 498 (1925)

[17] S. W. Hawking, Black Hole Explosions?, Nature **248**, 30 (1974)

[18] S. W. Hawking, Particle creation by black holes, Comm. Math. Phys. **43**, 199 (1975)

[19] W. G. Unruh, Experimental black-hole evaporation, Phys. Rev. Lett. **46**, 1351 (1981)

[20] C. Barceló, S. Liberati, M. Visser, Analogue gravity, Living Rev. Relativ. 14 (2011)

[21] J. Steinhauer, Observation of quantum Hawking radiation and its entanglement in an analogue black hole. Nature Physics, **12**, 959 (2016)

[22] R. Dardashti, S. Hartmann, K. P. Thébault, E. Winsberg, Hawking radiation and analogue experiments: A Bayesian analysis. Studies in History and Philosophy of Modern Physics **67**, 1 (2019)

[23] K. Crowther, N. Linneman, C. Wüthrich, What we cannot learn from analogue experiments, Synthese 198, Suppl. **16**, 3701 (2021)

[24] D. K. Buchwald et al (eds.), *The Collected Papers of Albert Einstein*, Vol. 14, Princeton (2015)

[25] E. Schrödinger, *Collected Papers on Wave Mechanics*, Minkowski Institute Press, Montreal, 2020

[26] M. Born, Quantenmechanik der Stossvorgange. Z. Phys. **38**, 803 (1926)

[27] E. Schrödinger, Über das Verhältnis der Heisenberg-Born-Jordanschen Quantenmechanik zu der meinen, Ann. Phys. **79**, 734 (1926)

[28] E. Schrödinger, Quantisierung als Eigenvertproblem, Ann. Phys. **79**, 361(1926)

[29] L. de Broglie, La mécanique ondulatoire et la structure atomique de la matière et du rayonnement, J. Phys. Rad. **8**, 225 (1927)

[30] L. de Broglie, Interpretation of quantum mechanics by the double solution theory, Ann. Fond. Louis de Broglie, **12**, 1 (1987)

[31] M. Jammer, *The Conceptual Development of Quantum Mechanics*, McGraw-Hill, New York (1966)

[32] F. Bloch, Heisenberg and the early days of quantum mechanics, Physics Today, Dec. 1976

[33] W. Moore, *A Life of Erwin Schrödinger*, Cambridge University Press (1994)

[34] L. de Broglie, *Non-linear Wave Mechanics*, Elsevier, Amsterdam (1960)

[35] G. Bacciagaluppi, A. Valentini, *Quantum theory at the crossroads: Reconsidering the 1927 Solvay Conference*. Cambridge University Press. Cambridge, 2009

[36] C. Davisson, L. H. Germer, Diffraction of Electrons by a Crystal of Nickel, Phys. Rev. **30**, 705 (1927)

[37] G. P. Thompson, Diffraction of Cathode Rays by a Thin Film, Nature **119**, 890 (1927)

[38] G. Galilei, *Siderius Nuncius*, Baglioni, Venice (1610), trans. by S. Drake in *Telescopes, Tides and Tactics*, University of Chicago Press, London (1983)

[39] Letter to Henslow, May 1860, in C. Darwin, *More Letters of Charles Darwin*, Vol. I, Francis Darwin and Albert Seward (eds.) Appleton and Co., New York (1903)

[40] J. Priestley, *The History and Present State of Electricity*, London (1767)

[41] P. Bartha, Analogy and Analogical Reasoning, The Stanford Encyclopedia of Philosophy (Fall 2024 Edition), E. N. Zalta and U. Nodelman (eds.)

[42] L. de Broglie, On the true ideas underlying wave mechanics, C. R. Acad. Sci. **B277**, 71 (1973)

[43] D. Shanahan, What might the matter wave be telling us of the nature of matter? International Journal of Quantum Foundations, **5**, 165 (2019)

Part II

RELATIVITY AND SPACETIME COSMOLOGY

Kyley Ewing (Ed.), *Spacetime Conference 2024. Selected peer-reviewed papers presented at the Seventh International Conference on the Nature and Ontology of Spacetime, 16 - 19 September 2024, Albena, Bulgaria* (Minkowski Institute Press, Montreal 2025). ISBN 978-1-998902-44-6 (softcover), ISBN 978-1-998902-45-3 (ebook).

4 Reinterpreting Relativity: Using the Equivalence Principle to Explain Away Cosmological Anomalies

Marcus Arvan

Abstract According to the standard interpretation of Einstein's field equations, gravity consists of mass-energy curving spacetime, and an additional physical force or entity—denoted by Λ (the "cosmological constant")—is responsible for the Universe's metric-expansion. Although General Relativity's direct predictions have otherwise been systematically confirmed, the dominant cosmological model thought to follow from it—the ΛCDM (Lambda cold dark matter) model of the Universe's history and composition—faces considerable challenges, including various observational anomalies and experimental failures to detect dark matter, dark energy, or inflation-field candidates. This paper shows that Einstein's Equivalence Principle entails two possible physical interpretations of General Relativity's field equations. Although the field equations facially appear to support the standard interpretation—that gravity consists of mass-energy curving spacetime—the field equations can be equivalently understood as holding that gravitational effects instead result from mass-energy accelerating the metric-expansion of a second-order spacetime fabric superimposed upon an absolute, first-order Euclidean space, resulting in the observational appearance of spacetime curvature. This alternative interpretation of relativity is shown to be empirically equivalent to the standard interpretation of relativity, albeit with a changing value for Λ (which is similar to how Λ is understood in the conception of Λ as "quintessence", but in this case takes Λ to be a feature of gravity). The reconceptualization is then shown to potentially resolve every major observational anomaly for the ΛCDM model, including recent observations conflicting with ΛCDM predictions, as well as failures to directly detect dark matter, dark energy, and inflation field/particle candidates.

> "[I]t is impossible to discover by experiment whether a given
> system of coordinates is accelerated, or whether its motion is straight
> and uniform and the observed effects are due to a gravitational field."
> Albert Einstein [20]

Physics is in crisis [2,43]. First, although the Standard Model of particle physics has been highly successful, it faces considerable theoretical [12, 65-6], explanatory [7, 13, 56, 69], and predictive [1, 14] difficulties. Second, the dominant theory of cosmology—the ΛCDM (Lambda cold dark matter) model of the Universe's composition and history [55]—faces equal if not more considerable challenges. Despite positing dark matter [71], dark energy [53, 70], and an inflation field [32-3] to account for a variety of observations, every experimental search for dark matter, dark-energy, and inflation-field candidates has thus far turned up empty [5]. Finally, recent observations of the cosmos appear to directly contradict the ΛCDM model. First, the Universe appears to be expanding faster than the ΛCDM predicts, suggesting that the Universe may be about 5 billion years younger than previously estimated using the

Kyley Ewing (Ed.), *Spacetime Conference 2024. Selected peer-reviewed papers presented at the Seventh International Conference on the Nature and Ontology of Spacetime, 16 - 19 September 2024, Albena, Bulgaria* (Minkowski Institute Press, Montreal 2025). ISBN 978-1-998902-44-6 (softcover), ISBN 978-1-998902-45-3 (ebook).

ΛCDM model [40, 58, 62-3]. Second, more recent observations have found an unexplained discrepancy between the observed expansion rate of the Universe just after the Big Bang and measurements in the local Universe, i.e. in nearby galaxies [64]. Third, recent observations of galaxies diverge from the predictions made by conventional models of dark matter [45]. Finally, recent images from the James Webb Space Telescope revealed a high number of high-redshift galaxies with unexpectedly high stellar masses [10]—a finding also in direct tension with the ΛCDM model.

In the past, similar crises in physics have been resolved not via more data collection [39, 65], but instead through what Thomas Kuhn famously termed "revolutionary science" [38]: that is, through paradigm shifts whereby relevant physical phenomena were reconceptualized—as in the cases of Copernicus reconceptualizing the planets as revolving around the Sun and Einstein reconceptualizing space and time as relative rather than absolute. Might physics be due for another paradigm shift? The present paper aims to show just this: that the physical significance of relativity's field equations may have been misinterpreted, and with it the physical constituents of the Universe as a whole.

1. Interpreting Einstein's Field Equations: Philosophical Preliminaries

Einstein's field equations are a set of ten equations that define gravitation in terms of the "curvature" of spacetime by mass and energy [24]. Here is one equation, the so-called "Einstein tensor":

$$G_{\mu\nu} = R_{\mu\nu} - \frac{1}{2}Rg_{\mu\nu}$$

Here is another:

$$G_{\mu\nu} + g_{\mu\nu}\Lambda = \frac{8\pi G}{c^4}T_{\mu\nu}$$

In these equations, "G" stands for Newton's gravitational constant, "R" stands for scalar curvature (the simplest non-Euclidean curvature in non-Euclidean Riemannian geometry), "$R_{\mu\nu}$" for the Ricci curvature tensor (viz. the amount by which the volume of a narrow conical piece of a geodesic ball in a Riemannian manifold deviates from that of the ball in Euclidean space), "Λ" for the cosmological constant, "$T\mu\nu$" for the energy momentum tensor (describing the density and flux of energy and momentum in spacetime), and "c" for the speed of light. Given that the field equations describe metric tensors in non-Euclidean spacetime, the most natural interpretation of their physical significance—the one presented by Einstein and now widely accepted [47]—is that they describe gravitation (viz. G) in terms of the density and flux of mass-energy curving spacetime (viz. $R_{\mu\nu}$). Nevertheless, dating back at least to Quine, philosophers have recognized that any term in a language (including equations) always admits of multiple interpretations. Indeed, Quine argues that there are always three related indeterminacies related to meaning. First, there is inscrutability of reference, or the fact that any sentence in a language can always be translated into a variety of other sentences referring to very different entities. To take a simple example, any coordinates in a non-Euclidean manifold can be translated into coordinates in Euclidean space (Fig 1).

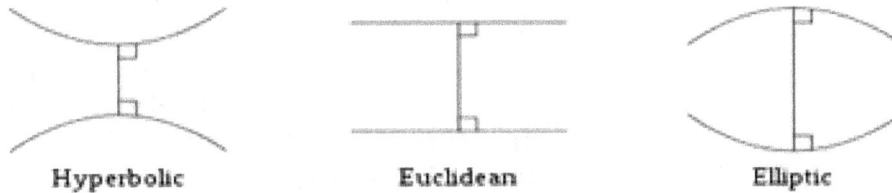

Figure 1: Euclidean 'Translations' of Non-Euclidean Geometry (fn.1)

Second, Quine argues that this in turn generates holophrastic indeterminacy, which is that empirically equivalent translations of sentences will nevertheless differ in terms of their ontological import—that is, in terms of what they posit to exist [59]. The present paper exploits these two indeterminacies as follows: we contend that following Einstein's Strong Equivalence Principle, relativity's field equations can be equivalently interpreted in two different ways: (A) the traditional interpretation (viz. mass-energy curving spacetime), or (B) in terms of mass-energy locally logarithmically accelerating the metric expansion of a second-order spacetime fabric around objects located in an absolute first-order Newtonian coordinate system—which we argue generates "spacetime curvature" as a measurement artifact. This brings us to one Quine's third indeterminacy, the underdetermination of scientific theory by empirical evidence [56,63]. Insofar as these different interpretations of relativity are equally consistent with observations taken to date, empirical data presently underdetermine which interpretation is more likely to be true. Which interpretation of relativity is more likely to be true can only be determined moving forward: by determining which interpretation generates better predictions in the future—such as our reconceptualization's predictions that (i) dark energy, dark matter, and inflation particles will never be discovered; (ii) the "Universe's" observed rate of expansion should continue to approximate a logarithmic function wherever it is observed, but (iii) the observed metric expansion of the Universe should appear to be different relative to different local gravitational systems.

2. Equivalently Reinterpreting Relativity's Field Equations

Let us return to the two relativistic field equations given earlier. According to their traditional interpretation, "G" is understood as standing for Newton's gravitational constant; "c" is understood as standing for the speed of light; and all of the other major terms besides the cosmological constant—"$T_{\mu\nu}$", "$g_{\mu\nu}$", and "$R_{\mu\nu}$"—stand for energy-momentum, metric, and curvature tensors, where tensors are (to simplify greatly) functions in coordinate space. So, if we set aside the cosmological constant, what these equations seem to say is that gravity is a function of the energy on objects generated by curved spacetime. Notice, however, that we have yet to interpret the cosmological constant ("Λ"). Einstein included this term because he saw that without it the Universe would collapse [23]. Einstein's inclusion of Λ is obviously justified, since the Universe hasn't collapsed. However, in the decades since Einstein introduced Λ, observations indicate that Universe's spacetime metric is not only not collapsing but instead expanding [36]. Consequently, theorists have supposed that "Λ" must refer to some yet-to-be-observed entity that causes spacetime to expand: either dark energy, a field of constant negative energy pressure, or quintessence, an entity akin to dark matter but the value of which changes over time [11,53,60]. Yet, although this substance is estimated to constitute about 70% of the Universe's mass-energy [27], no such substance has ever been directly detected in any experiment. Further, observational evidence of the cosmos has—at least on the traditional interpretation of the field equations—discovered another set of "anomalies." Galactic rotation curves [16], velocity dispersion profiles of elliptical galaxies [6], and galactic gravitational lensing effects [74] all suggest that the amount and distribution of mass-energy in different structures of the Universe are dramatically different than predictions suggest they should be given observed baryonic matter. These anomalies have led theorists to posit a second as-yet-detected substance [19]—dark matter—as constituting approximately 27% of the Universe's mass-energy [71]. Consequently, according to the standard interpretation of Einstein's field equations, our best theory of cosmology—the ΛCDM model—entails that between $95-97\%$ of the Universe's mass-energy is constituted by theoretical entities never directly confirmed in any experiment to date [27]. Further, these values not only appear to have changed dramatically across Universe's history; they appear to still be changing for yet-to-be understood reasons, as the Universe appears to be expanding more quickly than earlier observations and the ΛCDM model jointly predict it should, indicating that its "dark energy" is increasing [63]. Finally, it has been argued that a third as-yet undetected entity—an inflation field comprised by particles called "inflatons"—may be necessary to explain exponential expansion in the early Universe [32].

To see how reconceptualizing relativity may explain away these and other anomalies, let us begin with Einstein's strong equivalence principle [22]. In brief, this principle holds that that the force of gravity experienced by a person standing on a massive object is observationally equivalent to the force experienced by an observer in an accelerating frame of reference. Einstein [56] famously illustrated this equivalence through an example involving a person locked in a windowless elevator in outer space, noting that if the elevator were to accelerate upward, the person inside the elevator would experience themselves as "pulled" toward its floor by a seemingly invisible force. Consequently, Einstein concluded that the "downward pull" of gravity on Earth is empirically equivalent to the "upward" acceleration of a non-inertial reference frame. Allow us to now reinterpret this equivalence in a new way.

2.1. Gravity as Locally Accelerated Metric-Expansion of Second-Order Spacetime

Consider, to begin, two objects ("particles") located in absolute, unchanging Euclidean space (Fig 2).

Next, let us superimpose a second spatial metric—this time a dynamic (or changeable) spacetime fabric—on top of that first Euclidean space (Fig 3).

Let us imagine next the two particles described above as remaining precisely where they are in absolute first-order Euclidean space while making that absolute Euclidean space "invisible." We can do this, in pictorial form, by simply taking away the absolute Euclidean grid (Fig 4).

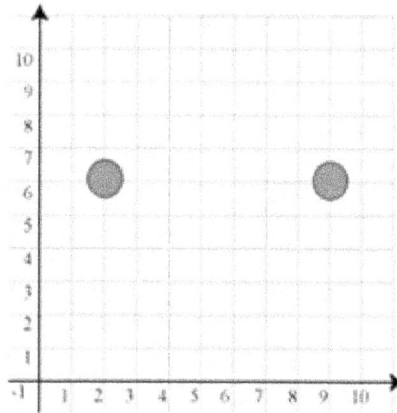

Figure 2: Two 'Particles' in Absolute Euclidean Space (fn. 2)

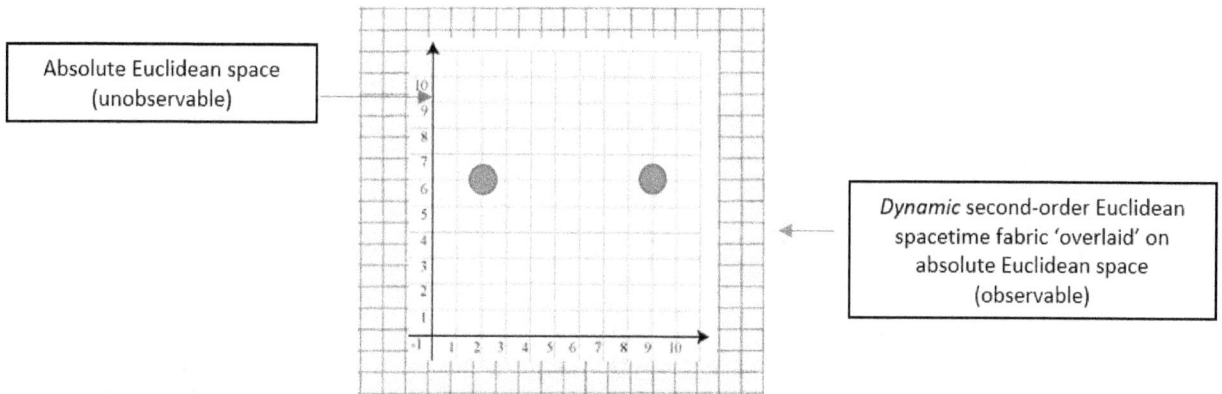

Absolute Euclidean space (unobservable)

Dynamic second-order Euclidean spacetime fabric 'overlaid' on absolute Euclidean space (observable)

Figure 3: Superposition of Dynamic Euclidean Spacetime Fabric on Absolute Euclidean Space

Remember, these two particles are now to be understood as located precisely where they were always located in absolute space. This new spatial grid is not a representation of absolute space, but now instead as a dynamic second-order fabric that surrounds those objects in first-order space.

Let us now consider "Λ." In contrast to the traditional interpretation of Λ, which understands it as something in addition to gravity, let us instead take Λ to refer to the locally accelerated expansion of the dynamic, second-order spacetime described above—while the two "particles" remain entirely unmoved from their previous locations in absolute first-order space. If we assume this, then observers in dynamic second-order space will observe the following (Fig 5):

Remember, the two particles pictured here have not moved at all from where they were located in the (now-invisible) first-order Euclidean space. Particle 1 has remained stationary at (2,6) in absolute space, and particle 2 has remained at (9,6). However, their spatial location in that first-order Euclidean space is invisible, as it is "beneath" the dynamic, second-order fabric those same particles are situated upon—the only spatial locations that observers in this world can observe. The point then is this: if we consider that space-the

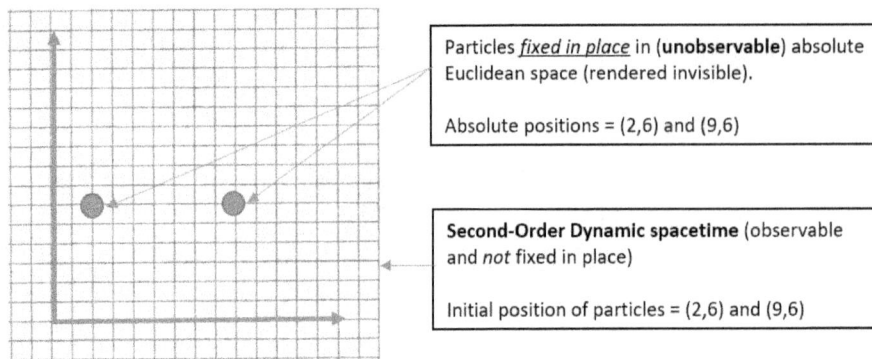

Figure 4: Two Objects Located Non-Observable Absolute Space Embedded in Dynamic Spacetime

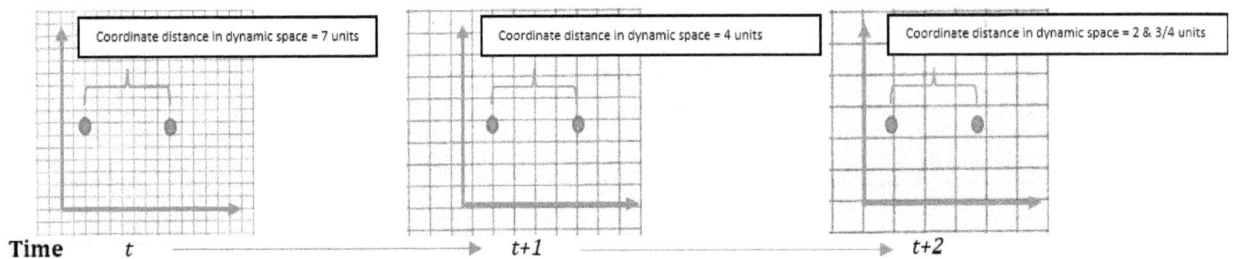

Figure 5: 'Gravitational Force' as Locally Accelerated Expansion of Dynamic Fabric

empirically detectable, expanding second-order space-then our observations will indicate that the two particles have "moved toward each other." At time t, the two particles were 7 observable spacetime units apart; at $t + 1$ they are 4 observable units apart, etc. (Fig 6).

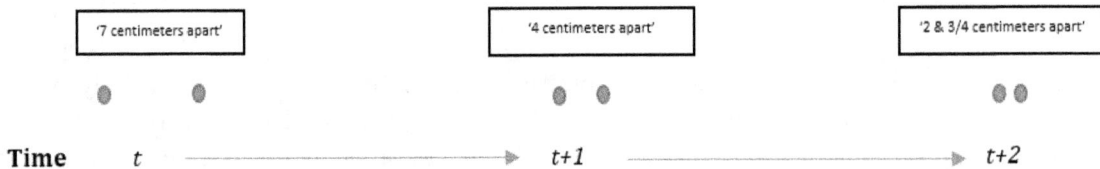

Figure 6: Measurements of object locations by observers in dynamic spacetime

Observers, in other words, will witness the two particles "drawing closer together" as if tugged toward each other by an invisible force—the force of gravity. So, our alternative interpretation of relativity models gravitational attraction. Yet, if this is the real mechanism of gravity, then for objects with mass-energy to continue accelerating toward each other at an increasing rate as they draw closer together (vis-à-vis Newton's Constant), the mechanism described above—objects with mass-energy expanding the local fabric of observable, dynamic spacetime—cannot occur at a constant rate. This is for the simple reason that as spacetime expands, the cubic volume of each unit of spacetime expands at an exponential rate. Consequently, for our reinterpretation of Einstein's field equations to correctly model gravitational behavior, the local expansion of second-order spacetime fabric around

76

objects with mass energy must increase the closer two objects of mass-energy get—that is, the expansion must accelerate to generate the Inverse-Square Law of gravitation (Fig 7).

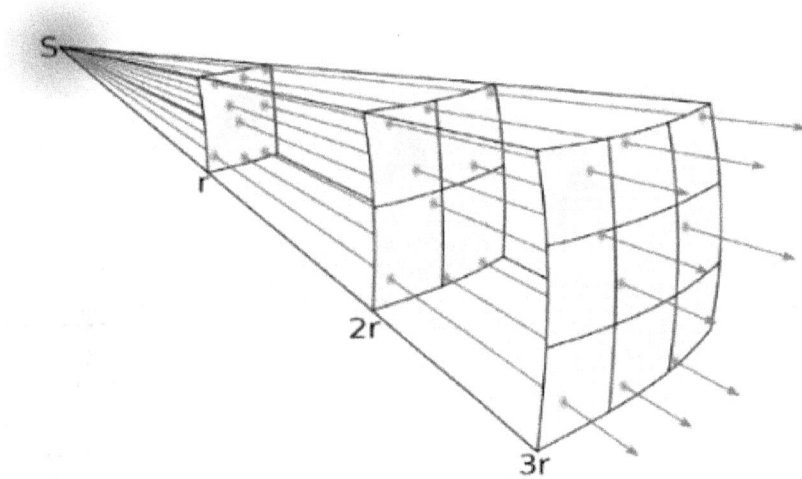

Figure 7: The Inverse-Square Law (fn. 3)

Next, let us plot a standard inverse-square function geometrically on a Cartesian plane (Fig. 8).

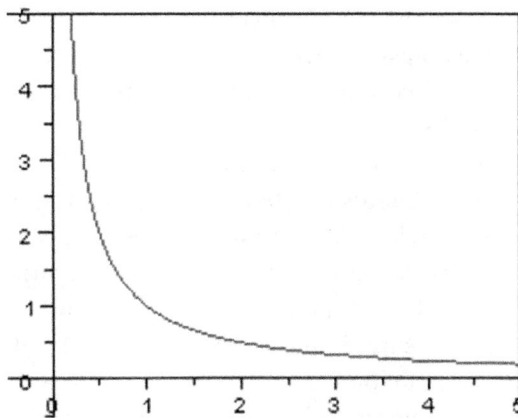

Figure 8: Inverse-Square Function (fn. 4)

Empirical observations indicate, again, that gravitational systems obey the Inverse Square Law, such that gravitational attraction is strongest the closer two objects of mass-energy are but weaken rapidly and progressively the further those objects are away from each other (viz. the inverse square of their distance). According to the traditional interpretation of relativity, this feature of gravitation results from how objects with mass-energy curve spacetime—viz. the intensity of gravitational curvature varying inversely to the square of distance from the gravitational source. Now, however, let us instead interpret the Inverse Square Law not in terms of curvature but instead in terms of accelerated expansion of the coordinate system surrounding a gravitational source across time (Fig 9), viz "Λ".

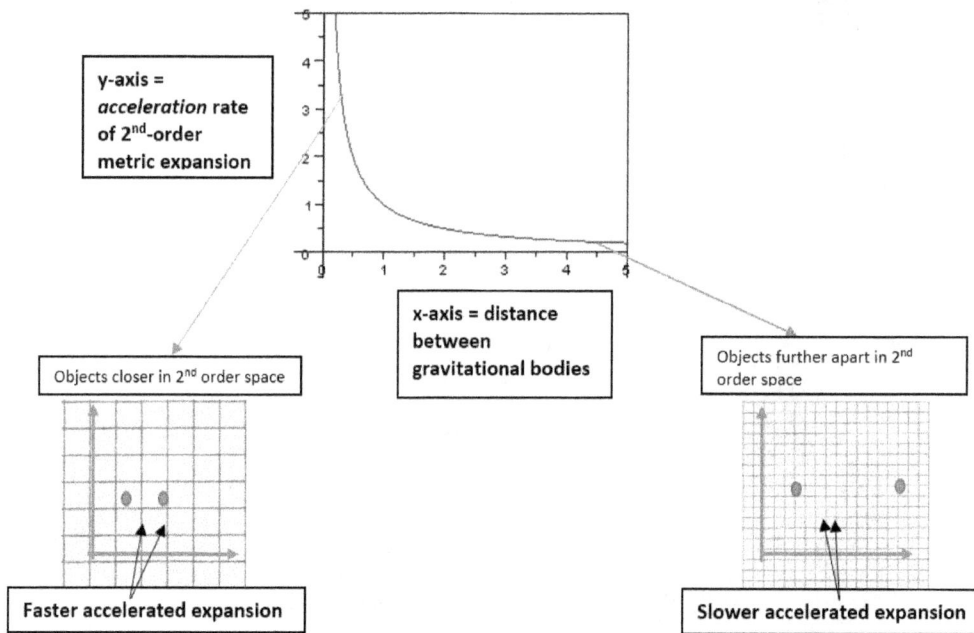

Figure 9: Interpreting Inverse Square Law as Accelerated 2nd-Order Metric Expansion

On this reinterpretation of relativity, the acceleration rate of the metric expansion of 2nd order space will be stronger the closer objects are to each other, viz. the Inverse Square function-accelerating metric expansion with the greatest intensity across time closest to the gravitational source, with that acceleration decreasing in inverse proportion to spacetime distance from the gravitational source.

As we will see below, this dynamic temporal behavior of second-order spacetime, the accelerated expansion of its coordinates across time by gravitational bodies (viz. the Inverse Square Law), will produce "gravitational curvature" as a measurement artifact. Before we do, however, it is critical to tease out two critical observational consequences of this interpretation. Notice, first, that if we understand gravitation as accelerated expansion of a second-order spacetime metric superimposed upon objects located within an unchanging first-order Newtonian metric, it follows that the cubic volume of space between any four coordinate points will grow exponentially larger. This is just a consequence of the geometry of cubic volume: the volume of a $1 \times 1 \times 1$ cube $1^3 = 1$, a $2 \times 2 \times 2$ cube is $2^3 = 8$, a $3 \times 3 \times 3$ cube is $3^3 = 24, 4 \times 4 = 4^3 = 64$, etc. Consequently, accelerated expansion of space by mass-energy (viz. gravity) should result in an exponential expansion of the volume of 2nd-order spacetime across time (Fig 10), "diluting" its volume exponentially as that spacetime expands.

To be clear, this exponential volume expansion will occur in direct proportion to how quickly expansion is accelerated by distance from a gravitational body, viz. the inverse-square law (as, following the inverse-square law, spacetime expansion will be accelerated the most near objects of mass-energy—thus exponentially increasing the volume of spacetime more the closer one is to a gravitational object—whereas the same exponential increase de-

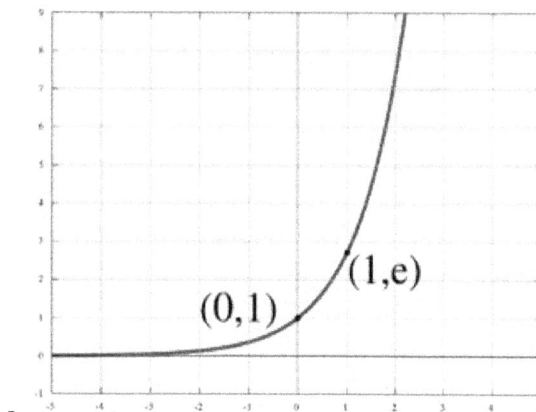

Figure 10: Exponential Function (fn. 5)

crease by the inverse square of distance). Thus, any gravitational system should—on our reconceptualization of the field equations—appear to exponentially accelerate the volume of 2nd -order space multiplied by the inverse square of distance. Now, following Einstein's equivalence principle, let us investigate what the observational consequences of this exponential volume expansion should be in a gravitational system. Just as Einstein's elevator accelerating upwards will generate the experience of "being pulled downwards on anyone in the elevator, an exponential function applied to spacetime coordinates should generate observable consequences equivalent to its inverse function: namely, a logarithmic function (Figure 11).

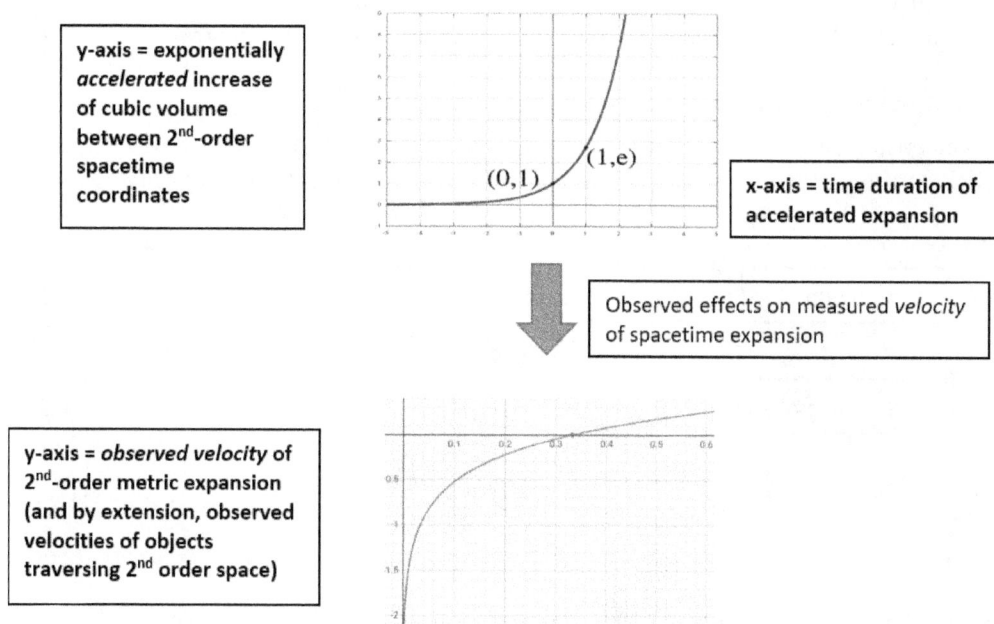

Figure 11: Equivalence of Exponential Volume Expansion to Logarithmic Behavior in Coordinate Space

The next thing to point out here is that, following the mathematics of exponential functions, these logarithmic consequences should be less apparent at small-scale levels and progressively more apparent in larger systems that accelerate the expansion of 2nd order spacetime more strongly due to having mass-energy. To see why, consider the mathematics of exponentiation for an exponential series using small numbers. $1^2 = 1, 2^2 = 4$, and $4^2 = 16$. As we exponentiate with small numbers, we get relatively small absolute increases in volume. Yet, now consider that 1 million squared is 1 trillion and 1 trillion squared is 10^{24}. The larger the base number that we begin with in exponentiation—which, in a gravitational system, will be determined by the mass-energy of the system—the more pronounced the observational implications of its exponential operations will be on the behavior of the system. And, of course, here we have only been dealing with squares, whereas in three-dimensional space we will be dealing with exponential increases in cubic volume. The differences between how exponentiation works on small base numbers compared to larger ones is critical, as it enables our reconceptualization to explain apparent divergences between gravitational behavior in smaller systems (e.g., the Solar System) and vastly larger systems (galaxies and the Universe).

2.2. 'Spacetime curvature' as observational artifact of accelerated spacetime expansion

Let us begin by considering a beam of light propagating by a massive object in spacetime, say the Sun. Following the reconceptualization of relativity sketched above, we are now to suppose that the Sun's mass-energy accelerates the expansion of 2nd-order spacetime fabric near the Sun via the Inverse Square function while the light propagates through unchanging 1st-order space (**N.B.**: the plots that follow do not follow an inverse square function, and so should not be taken to be physically realistic. The diagrams are merely to show conceptually this paper's reconceptualization of relativity explains spacetime curvature). Our reconceptualization entails that a massive object (such as a star) will exponentially expand the cubic volume of spacetime fabric around a propagating light wave (Fig 12).

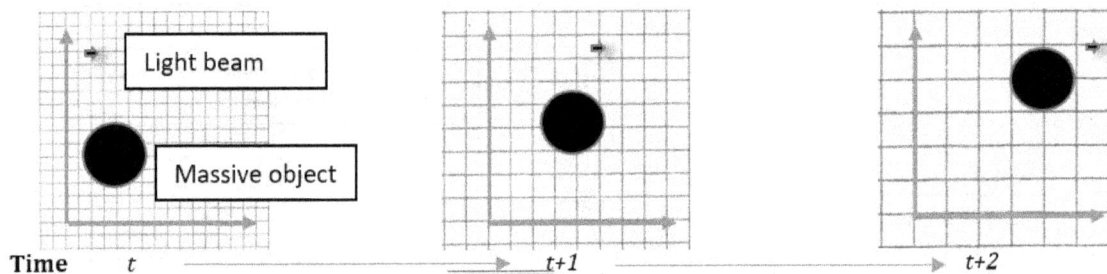

Figure 12: Light Traveling Straight Through Accelerated Expansion of Dynamic Spacetime Fabric

Notice what is happening here. At time t, the light beam will be measured by observers to be located at $(2, 13)$; at $t+1$, it will be observed to be coordinates $(5, 7.5)$; and at $t+2$, at $(5.5, 5)$, as in Fig 13. The path the propagating light beam will trace through the accelerating expansion of cubic spacetime volume across time will thus be curved relative to the center of mass. The "curved" path of the beam will thus be an artifact not of curved spacetime but instead mass-energy dynamically expanding spacetime in an exponential manner, resulting in measurements in dynamically expanding spacetime that realize a curved path through that spacetime fabric, which again is expanding exponentially across time in a way that results in inverse-square law-like behavior (as the inverse effect of expansion on the object's path).

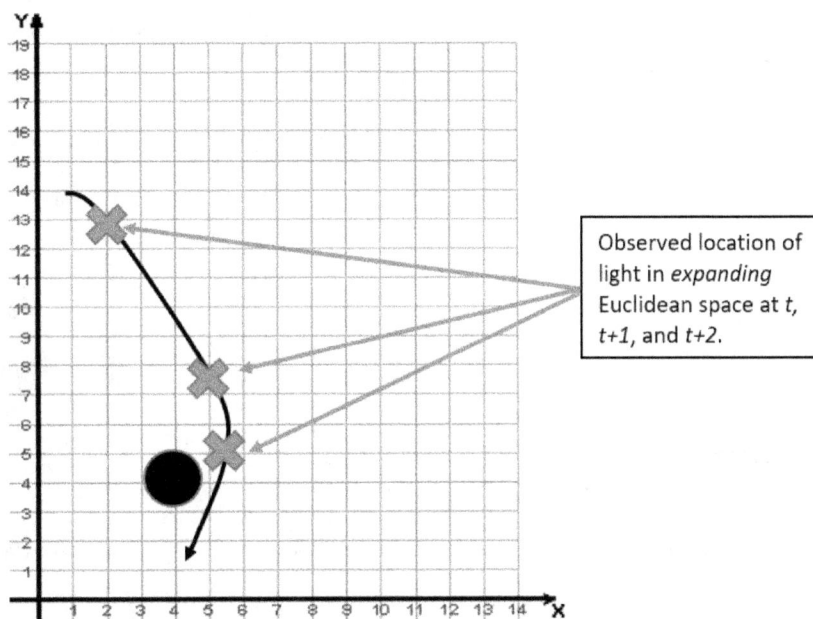

Figure 13: Gravitational Curvature as Artifact of Λ as Local 2nd-Order Accelerated Metric Expansion

Consequently, reinterpreting relativity in terms of the accelerated local expansion of spacetime fabric by mass-energy leads to observations of the apparent curvature of space by gravity. On this interpretation of the field equations, the apparent curvature of spacetime is simply an observational artifact of mass-energy causing the accelerated expansion of a second-order Euclidean spacetime.

As we will see shortly, this interpretation of relativity is capable of systematically explaining away anomalies generated in the ΛCDM model by the traditional interpretation of the field equations. But first, notice next that because light is a particle and a wave, this reinterpretation implies that as light travels through a gravitational field its wavelength will appear stretched, qua "redshift" (Fig 14).

Now let us turn to the observed speed of ordinary objects. Let us begin by plotting the spatial position of an object over time in (unobservable) absolute Euclidean space. Let us suppose, specifically, that this object is me walking from one place to another at a constant rate relative to absolute space, e.g. 2 spatial units per 1 unit of "objective" time. This path through Euclidean space is represented in Fig 15.

Next, let us suppose that I am walking on an object (the Earth) with a high mass-energy. Consequently, on the reinterpretation of relativity being proposed, the time that the

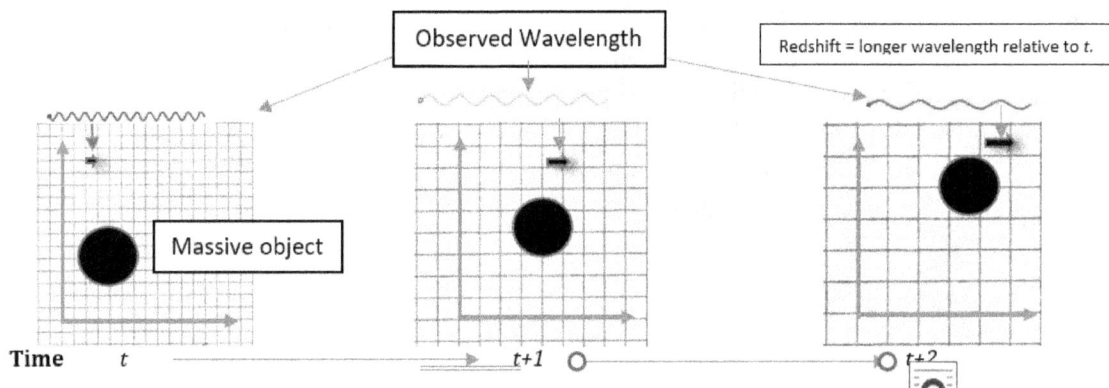

Figure 14: Redshift = Gravity as Locally Accelerated Expansion of 2nd Order Euclidean Spacetime (fn. 6)

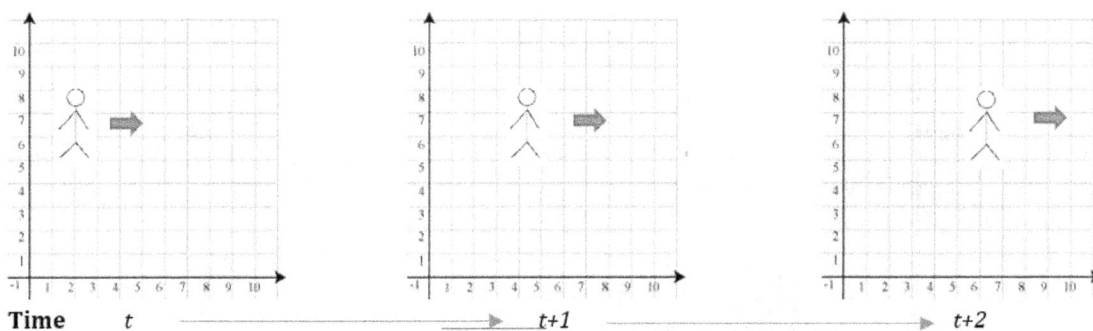

Figure 15: Movement of Object Through Absolute (Unobservable) Euclidean Space (fn. 7)

moving object takes to traverse spacetime near a massive object will expand, slowing down relative to other spacetime coordinates not near the massive object (viz. "time-dilation"). Because cubic spacetime is expanding around objects, objects in a gravitational field will take progressively longer and longer to traverse a single unit of spacetime as time evolves (Fig 16).

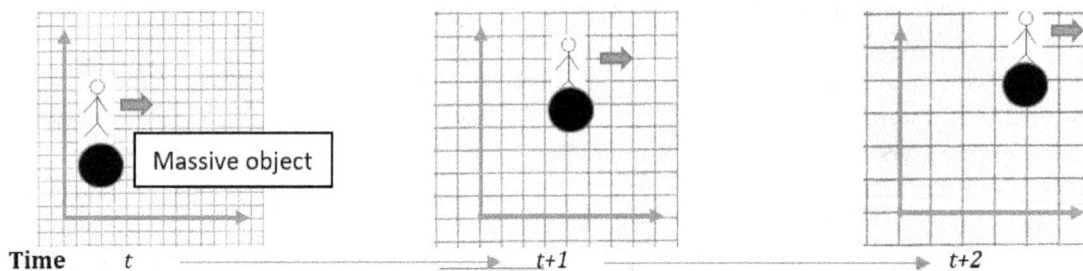

Figure 16: 'Time Dilation' as Consequence of Accelerating Metric-Expansion of Euclidean space

Observers within this expanding Euclidean space (i.e. you and me) would witness nothing odd: we would experience ourselves as moving at a constant rate—as everything in this reference frame would slow down at the same rate. However, to observers outside of our gravitational reference-frame, things would appear very different. Because their spacetime would not be caught in the local expansion of our gravitational field, they would witness everything in our vicinity taking longer and longer to occur. That is, relative to outside observers, the gravity surrounding us would appear to slow time down, as they would see individuals taking longer and longer to traverse a single observable spacetime metric. The reconceptualization of Einstein's field equations being proposed thus explains "time dilation": it just does so via a different mechanism the traditional interpretation of the field equations posits.

2.3. 'Dark Energy' and 'Dark Matter' as Observational Artifacts of Logarithmic Expansion

The ΛCDM model holds that the Universe is constituted by three things:

1. Ordinary "baryonic" matter and energy (quarks, atoms, electromagnetism, etc.)

2. A cosmological constant (Λ) associated with dark energy, a special kind of energy that is thought to accelerate the metric expansion of the Universe equally throughout all space.

3. Cold dark matter (CDM), a special type of matter that moves very slowly and has gravitational effects but interacts very weakly with ordinary matter and electromagnetic radiation.

This model has been arrived at based on the traditional interpretation of relativity's field equations along with a variety of observations of the cosmos. First, because observations of the cosmos suggest that spacetime is expanding [36], theorists have supposed that "Λ" has to stand for some additional theoretical entity beyond gravity: either dark energy, an unseen force that expands spacetime throughout the Universe at a constant rate of acceleration, or "quintessence", an unseen force that expands the Universe at a variable rate. Second, cosmological observations suggest—on the traditional interpretation of relativity—that galaxies are surrounded by vast haloes of unexplained mass. Evidence for this "extra mass" comes in several forms. First, spiral galaxies have unexpectedly "flat" rotation curves [16]. Second, the rotation curves of galaxies appear to have changed dramatically from the early Universe to today [25,29]. These observations have been taken by theorists to imply that early galaxies were dominated by ordinary matter, only to become more dominated by dark matter as the Universe has progressed [26]. Third, velocity dispersions (the rate at which objects at different distances move) in elliptical galaxies do not match predictions based on those galaxies' observed ordinary baryonic matter [7]. Fourth, galaxies in general have much stronger gravitational lensing effects (the amount that they bend starlight) than predicted using observations of their ordinary baryonic matter [61,73]. Finally, there is at least one further oddity that lacks any explanation on the ΛCDM model. Recent observations indicate a "strange relationship" between galactic supermassive black holes and dark matter [8-9]: namely, that "the more dark matter a galaxy has, the bigger its black hole tends to be" [48].

Crucially, all of these "anomalies" are based upon the traditional interpretation of General Relativity's field equations. But let us not mince words at this point: the ΛCDM mode is a theoretical and predictive mess. First, the ΛCDM model posits not one but two (and perhaps three) theoretical entities—dark matter, dark energy, and a distinct inflation field—that have never been directly observed in any experiment. Second, the ΛCDM model holds that the amounts of different forms of mass and energy—including the proportions of dark energy, dark matter, and ordinary baryonic matter—have changed dramatically over the course of the Universe's history for reasons that no one understands [26]. Third, estimations of the Universe's rate of expansion based on the ΛCDM model and previous observations directly conflict with the rate of expansion found in more recent observations [63]. Fourth, even more recent observations [64] suggest that the local Universe is expanding more quickly than observations of the distant Universe just after the Big Bang suggest it should, assuming the ΛCDM model is true and that there are more massive galaxies in the early Universe than the ΛCDM predicts [10]. Fifth, the ΛCDM model contains no obvious explanation of why galaxies with larger central black holes should have more dark matter. Sixth, dark matter simulations indicate that the density of dark matter should be more "peaked" in galaxies than observed [28]. Finally, the ΛCDM model provides no clear explanation for exponential inflation in the early Universe. We could go on—but the point is this: if you wanted to design a false scientific paradigm akin to Ptolemy's epicycles or the luminiferous aether, you could hardly do better than this.

We have already seen how this paper's reconceptualization of relativity can explain observations of "spacetime curvature", time-dilation, and redshift of light from the distant Universe. We will now see that it can explain away dark energy, dark matter, the inflationary epoch of the early Universe, and why the Hubble constant appears to have different values in the local and early Universe. Let us begin with dark energy. The current paradigm—embodied in the ΛCDM model—is that "Λ" in the field equations stands for a constant in nature: a repulsive force that is expanding the Universe's spacetime metric everywhere at a constant rate. The main evidence for this account has been observational data indicating a linear relationship between the distance between us and observed galaxies and those galaxies' redshift [62-3]. Further, the idea that space is expanding in this uniform fashion has been codified in what is known as Hubble's Law (recently renamed the Hubble-Lemaître Law [37]). Alas, as we have just seen, there is a problem here: a variety of recent observations with the Hubble Space Telescope indicate that the Hubble Law is false. First, observations indicate that the Universe is expanding significantly faster than predicted using the ΛCDM model and previous observations [52, 62]. Second, galaxies in particular clusters (e.g. the Virgo cluster) deviate significantly from the otherwise linear relationship between distance and redshift posited by Hubble's Law [68]—see Fig 17. Third, as noted above, local observations suggest that the Universe is expanding at a different velocity than distant observations [64].

On the ΛCDM model and traditional interpretation of relativity, all of these findings are complete mysteries [44]. And indeed, some have already suggested that this unexplained deviation from the Hubble Law may require revisions to physics or to the ΛCDM model [71]. Yet, these results are not a mystery on the reinterpretation of relativity proposed herein. First, on our reinterpretation of relativity, the Universe's spacetime metric is not expanding everywhere at a uniform rate. Instead, spacetime expansion occurs locally—around

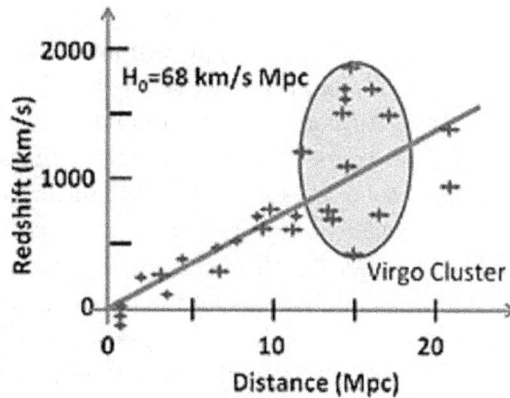

Figure 17: Deviations of Virgo Cluster Galaxies from Hubble's Law (fn. 8)

objects with mass-energy (i.e. planets, stars, galaxies, etc.). Second, our reinterpretation of the field equations thus explains these and other redshift deviations from Hubble's Law straightforwardly: as a local effect of gravitational systems on their surrounding spacetime coordinates according to a logarithmic function. According to this reinterpretation relativity, gravitational systems such as galaxies should have broadly similar redshift profiles that increase with age and distance (as Hubble's Law indicates), but nevertheless differ case-by-case depending upon (i) how much mass-energy the system has, (ii) how that mass-energy is distributed in the galaxy, and (iii) how long that system's mass-energy has been accelerating the expansion of its local spacetime metric (viz. the age of the particular galaxy itself). Third, our reinterpretation explains why the apparent amount of "dark energy" is orders of magnitude greater than the amount of ordinary baryonic matter observed in the Universe, and why its quantity appears to have changed dramatically over the course of the Universe's history. For, our reinterpretation holds equates gravity itself with mass-energy locally accelerating the metric-expansion of spacetime at a logarithmically accelerating rate-a rate that increases dramatically near the boundary of a system of mass-energy before rapidly "flattening off" but continuing to rise across time. We will illustrate this shortly.

Before we do—and the reasons for this will become clear later—let us now consider dark matter. Dark matter is thought to exist because, on the traditional interpretation of General Relativity, galaxies appear to have vastly stronger gravitational effects—viz. gravitational lensing, spiral rotation curves, and so on—than their visible matter suggests. The only way to explain this, on the traditional interpretation of General Relativity, is to hold that there is something—something that cannot be "seen" like ordinary baryonic matter (viz. interacting with electromagnetism)—giving those galaxies extra mass. The new interpretation of General Relativity outlined above promises to elegantly explain the above phenomena without positing dark matter. Here is how.

Consider first the Solar System. The Sun's mass is 1.989×10^{30} kg, constituting 99.8 percent of the Solar System's total mass. The Sun's diameter is 1.391 million km. The Solar System's diameter is 149,597,870 km. So, the Sun's diameter constitutes approximately 9.3% of the Solar System's diameter while containing nearly all of the Solar System's mass. Now consider the Milky Way galaxy. On April 20th, 2019 scientists released the first confirmed image of a black hole: an image of the supermassive black hole at the center of our own Milky Way galaxy, Sagittarius A* [50,54]. Observational estimates indicate that Sagittarius A*'s diameter is about 60 million km [42], and its mass between 3.7 ± 0.2 million and 4.31 ± 0.38 million solar masses [30 − 31]. In contrast, the diameter of the Milky Way Galaxy is estimated to be 150-200,000 light years [41], and the total mass of its ordinary baryonic matter approximately 60 billion solar masses [18]. This means that the center of gravity in our Galaxy—the supermassive black hole at its center—constitutes only .00006% of the galaxy's baryonic mass and only .000000012% of the galaxy's diameter. This means that the distribution of ordinary matter in solar systems and in galaxies are vastly different in orders of magnitude. In solar systems, ordinary baryonic mass-energy is around 98% centrally located (in the star at the solar system's center), whereas in galaxies the ordinary baryonic mass-energy is not centrally located, but instead more widely distributed throughout the galaxy.

As we saw earlier, on the reinterpretation of relativity being proposed, these differences in scale and mass-energy distribution matter when it comes to the functional characteristics of the system. For consider again the idea that gravitation is not a matter of actual spacetime curvature but instead perceived curvature by objects of mass-energy accelerating local metric expansion of 2nd-order spacetime fabric by the Inverse Square Law. As we saw in Figures 10 and 11, this accelerated expansion should exponentially increase the cubic volume of 2nd-order spacetime by mass-energy multiplied by the inverse square of distance from a gravitational object. As we saw earlier, since exponential functions result in vastly higher increases when applied to larger base numbers compared to smaller ones, it follows, via Einstein's equivalence principle, that objects located and moving in 2nd-order space should experience effects approximating the inverse of an exponential function, where this effect is far less (or even negligible) in small systems but pronounced in larger systems.

The implications here are straightforward. Objects in smaller gravitational systems should appear to obey the inverse-square law viz. their observed velocity through space-time. Yet, much larger gravitational systems that have been around longer (such as nearby galaxies) should appear to have more gravity over time—and flat rotation speeds indicative of "dark matter"—since exponential increases on the cubic volume of their local spacetime has been operating on a higher base number (high mass-energy) over a much longer period of time. But this is exactly what is observed and is otherwise a cosmological mystery: nearby galaxies do appear to have more "dark matter" than more distant (younger) galaxies, and nearby galaxies' rotation curves are logarithmic (Fig 18).

What about gravitational lensing? As we saw earlier, the bending of light is—on the new interpretation of the field equations we are proposing—a function of mass-energy locally accelerating the local expansion of dynamic spacetime at an increasing rate. In a mass-energy system like the solar system—where mass is centrally located—the metric expansion of space occurs primary toward the center of gravity, weakening dramatically the further out one moves away from the central source of gravity. However, according to the interpretation's analysis of galactic gravitation, spacetime expansion is accelerating across a much wider area over a longer period of time: the entire area of the galaxy. Because on the reconceptualization of relativity being offered, gravity is the accelerated local expansion of space by mass-energy, galaxies should appear to have "more gravity" than their observed baryonic matter

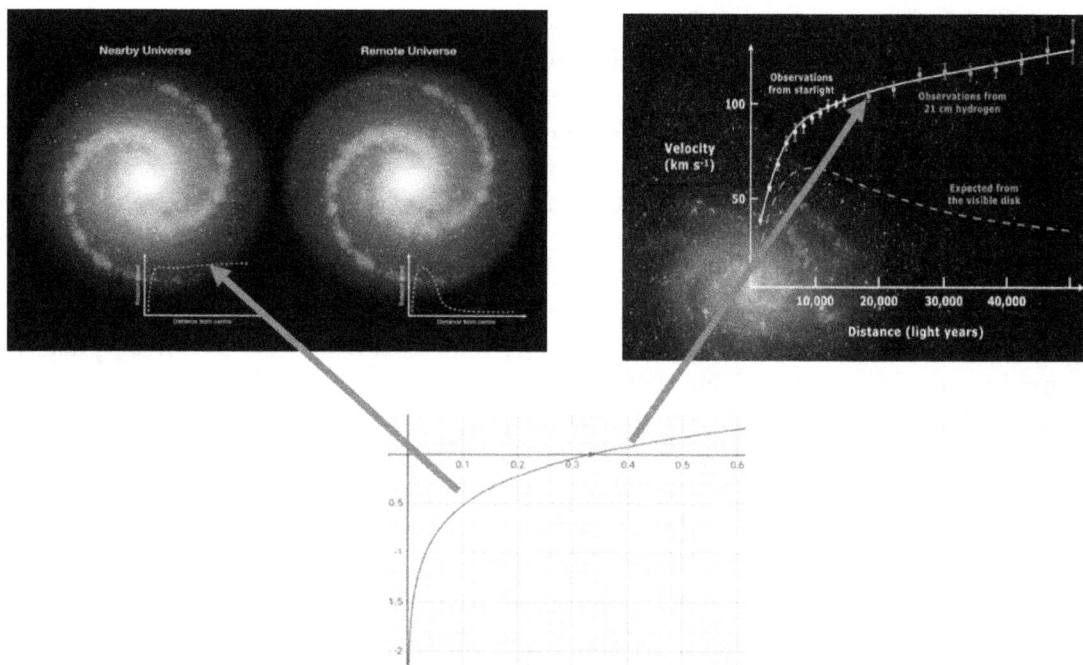

Figure 18: Reconceptualized Relativity Explains 'Dark Matter' Galactic Rotation Velocity Curves (fn. 9)

suggests—which is what gravitational lensing observations currently interpreted as "dark matter" indicate.

On this interpretation of "dark matter", there is obviously a direct connection to "dark energy": they are two sides of one and the same thing, namely gravitational effects being the result of mass-energy locally accelerating the expansion of spacetime. This unified explanation of the appearance of dark matter and dark energy not only explains them away (without us having to posit any such extra entities); it also explains some astonishing and otherwise unexplained coincidences. First, it explains why galaxies with larger supermassive black holes appear to have more "dark matter" [48]. Dark matter is nothing but gravity (properly interpreted according to a changing value for Λ), and galaxies with larger supermassive black holes have more gravity. Second, our reconceptualization explains another fascinating "coincidence" that arises in the mathematics of Modified Newtonian Dynamics (MOND). In brief, MOND holds that gravity operates differently in slowly accelerating systems like galaxies—where it holds that instead of varying inversely with the square of radius distance, gravity varies inversely simply with radius. There are many outstanding issues with MOND that we need not concern ourselves with here. Let us instead consider a few basic points. Here is MOND's central equation [46]:

$$\vec{F} = m\mu\left(\frac{a}{a_0}\right)\vec{a}$$

In this equation, F is Newtonian force, m is mass, a is acceleration, μ is an "interpolating" function, and a_a a new fundamental constant of nature demarcating the transition between Newtonian and MOND gravity. In other words, this equation describes how gravity

87

(supposedly) operates totally differently in conditions of low acceleration. Of most interest to us here is a_0. When a_0 is fit to the observed properties of galaxies, its value turns out to be within an order of magnitude of cH_0, where c is the speed of light and H_0 is the Hubble constant. In other words, MOND's equations demonstrate that—at least on its alternative theory of gravity—the altered properties of gravity in galaxies is approximately identical in value to the acceleration rate of the universe (viz. Λ). MOND does not provide any account of why this should be so, and as we have seen the standard interpretation of General Relativity does not explain this fascinating coincidence either. This paper's alternative interpretation of the field equations, on the other hand, explains it directly: the observed accelerated metric-expansion of the Universe (Λ) just is gravity, and the strange behavior of gravity on galactic scales (which MOND attempts to describe without dark matter) just is the consequence of the value that Λ must take on our new interpretation (an exponential expansion function).

2.4. Cosmic Inflation as Gravitational Effects of the Big Bang Singularity

Finally, this paper's reinterpretation of Einstein's field equations may even explain cosmic inflation. Currently, the dominant theory of the Universe's history holds that our Universe began from an infinitely dense point (i.e. the Big Bang). Following Hawking, who demonstrated that a Big Bang is mathematically equivalent to a time-reversed black hole [35], the Big Bang has been theorized to be a "white hole" [20]. Let us now think what this means. The only properties that a black hole has are mass, spin, and charge. If the Universe is a time-reversed black hole (i.e. a white hole), then the Big Bang singularity itself has an immense (potentially infinite) mass-energy. Consequently, if as the present paper has argued gravity itself is mass-energy accelerating the metric-expansion of second-order spacetime—which as we have seen has logarithmic velocity effects—then the increase in velocity of spacetime expansion should be immense just after the Big Bang before flattening off-just as cosmological observations indicate. Could this—that is, gravity itself—be the right explanation of exponential inflation in the early Universe (rather than some additional "inflationary field")? One hint that it may be is the fact that all of the "exotic" phenomena posited by the ΛCDM model-dark matter, dark energy, and the inflationary epoch of the Universe-correspond to logarithmic curves. Another hint is that recent observations of galaxies—which have "amazed" researchers—indicate that galaxy rotation speeds, while "flat", do not match conventional models of mass distributions and dark matter [15-6]. Instead, galaxy rotation speeds have been found to be highly correlated with their ordinary visible matter [45]—just as our reinterpretation predicts. It should not be underestimated just how much these recent findings confound dark matter theory, with researchers stating, "It's an impressive demonstration of something, but I don't know what that something is" [15]. This paper's reinterpretation of relativity explains them.

First, consider galactic rotation curves currently taken as evidence of dark matter—which correspond to the logarithmic curve generated by reinterpreting gravity as the exponential expansion of cubic spacetime volume (Fig 19).

Now compare this to the expansion curve of the Universe-currently thought to be explained in the early Universe by an inflationary particle/force and later by dark matter and

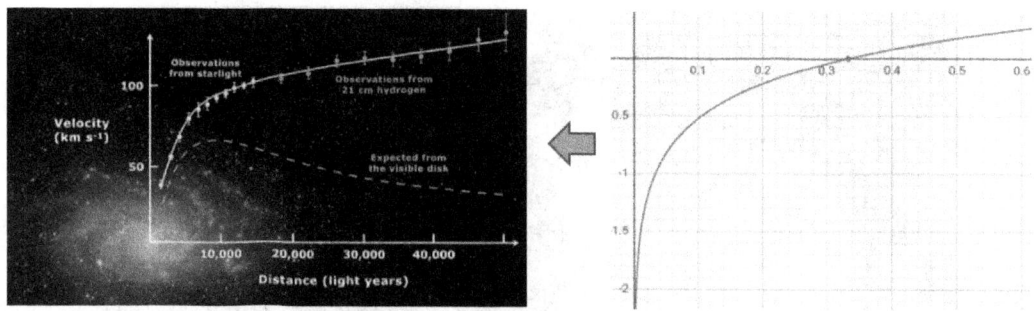

Figure 19: Fit of Logarithmic Velocity Implications of Gravity as Accelerated Local Metric Expansion to Nearby Galactic Rotation Curves

energy (Fig 20).

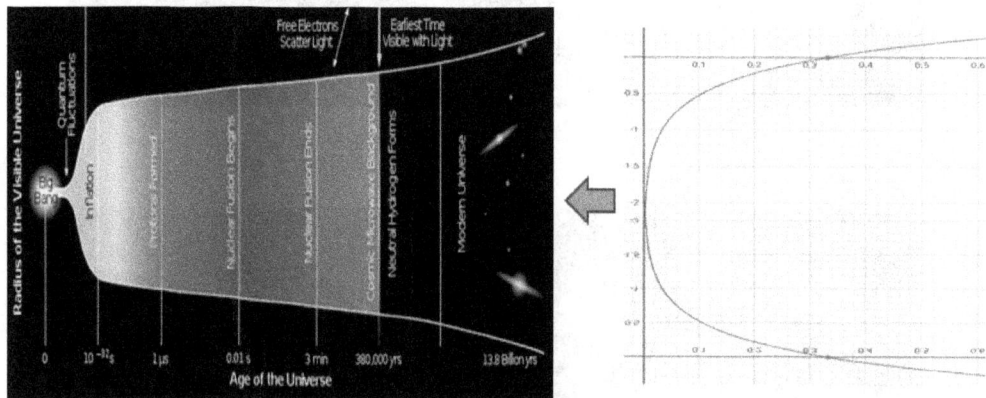

Figure 20: Fit of Gravity as Logarithmically Accelerated Local Metric Expansion to Inflationary Universe Curve (fn. 10)

The prevailing model of the Universe based on the ΛCDM model provides no explanation of these apparently stunning set of coincidences: (i) the functional properties of galaxies associated with "dark matter" (or modified gravity in MOND) is approximately identical in value to the acceleration rate of the universe (viz. Λ), both of which (ii) approximate logarithmic curves (Fig 21).

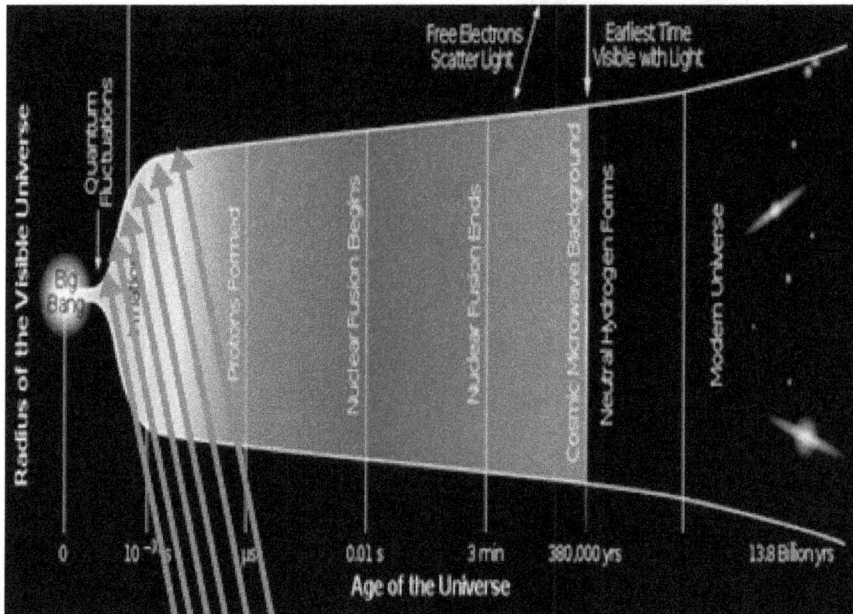

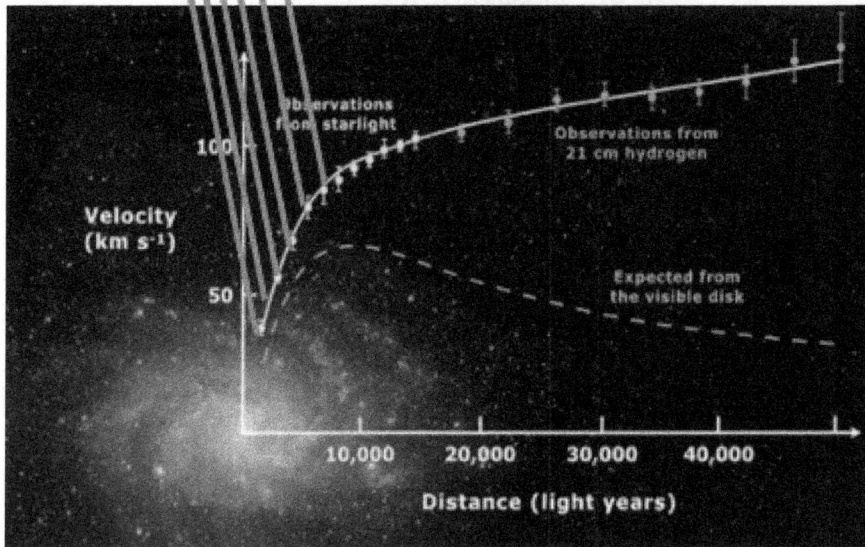

Figure 21: Unexplained Cosmic Coincidences in the ΛCDM Model

Further, the ΛCDM model provides no explanation for why the expansion rate of the Universe has evolved the way it has, or why (according to the ΛCDM model) the constituents of the Universe itself supposedly responsible for this behavior—and the changing behaviors of galaxies—have changed dramatically over the course of the Universe's history [26]—see Fig 22.

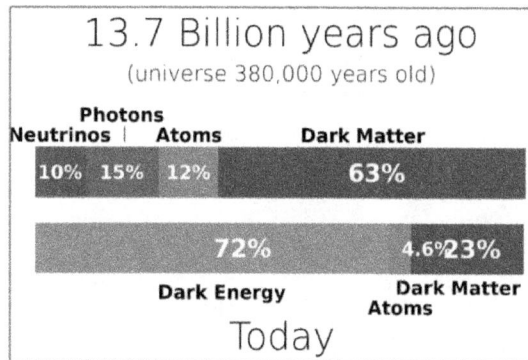

Figure 22: Changes in Universe's Hypothesized Composition Over Time According to the ΛCDM Model (fn. 11)

The ΛCDM model also fails to explain why estimates of the Universe's composition based on precision astronomical measurements taken before 2013 had to be significantly revised on based on updated 2013 measurements—and hence, if the measurements are correct (as they appear to be), why the Universe's composition still appears to be changing [63]. And, of course, we need to remember that the various supposedly responsible for these explained phenomena—"dark matter", "dark energy", and potentially a distinct inflation field in the early Universe comprised by particles such as "inflatons"—have never been directly detected in any experiment to date. Our reconceptualization of relativity explains all of the above coincidences and anomalies. We can see how by examining different hypothesized values for "Λ" on the ΛCDM model, and by extension hypotheses for outcomes of the future of the Universe. Currently, there are several open hypotheses about the value of Λ and fate of the Universe: the Big Rip Hypothesis that average density of the Universe (viz. gravity and "Λ") is such that spacetime expansion will accelerate, ripping spacetime part; the Continual Expansion hypothesis that the critical density of the Universe will result in a "flat" expansion, expanding eternally; and the Big Crunch Hypothesis that the Universe will ultimately contract due to gravity (Fig 23).

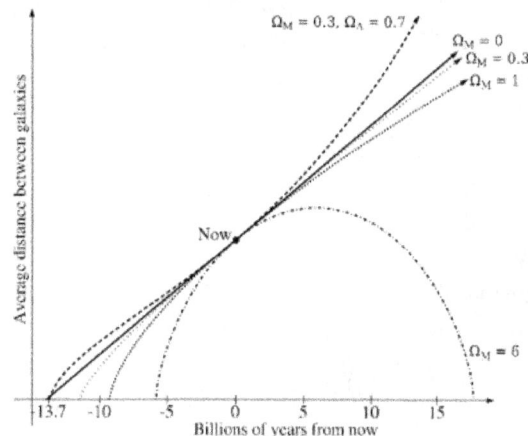

Figure 23: Possible values for 'Λ' and the fate of the Universe (fn. 12)

Crucially, however, the preponderance of evidence at the present time is that the Universe's expansion is flat [51]. But now what function explains and predicts this finding? Answer: A logarithmic curve—which becomes progressively flatter the further the function is extended in time (Fig 24).

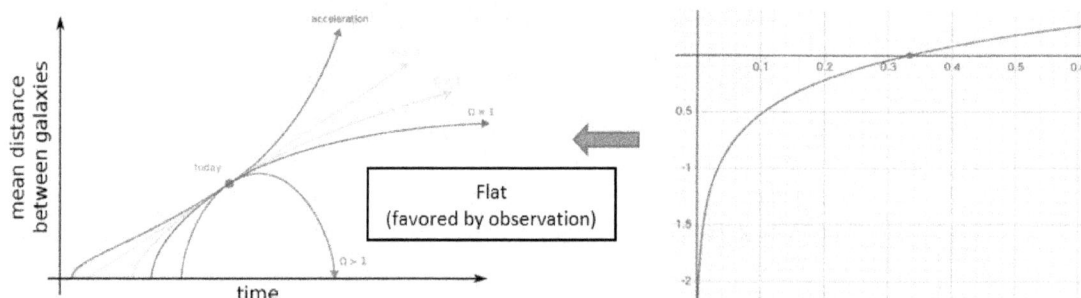

Figure 24: Flat Universe Predicted and Explained by Reinterpretation of Relativity (fn. 13)

But in that case, notice: this paper's reinterpretation of relativity predicts what has in fact been observed—the early exponential increase and progressive flattening of the Universe's velocity of expansion. Further, this is not only a unique prediction—one that, if we are correct, is justified by the general theory of relativity; it is also a finding that makes predictions about the future: namely, that increases in the Universe's measured expansion velocity from here on out should continue to progressively flatten, qua logarithmic curve.

Next, our account explains certain mysteries about the concept of a "white hole." Currently, it is not well understood how a white hole can occur in nature, as white holes are thought (following Hawking) to be equivalent to time-inverted black holes—leading, obviously, to questions about how time can become inverted in a way that mass-energy can escape the extreme gravitational effects of a singularity. On our reconceptualization of relativity, these problems evaporate. White holes are not time-reversed black holes. The assumption that they are time-reversed is based upon the background hypothesis that the Big Bang expanded spacetime whereas black holes (qua the standard interpretation of gravity) are thought to contract spacetime. On our reinterpretation of relativity, this seeming asymmetry is based upon a conceptual mistake: namely, the failure to see that gravity just is the accelerated local metric-expansion of space-time by mass energy. On our account, the Big Bang and ordinary black holes (including supermassive black holes) are fundamentally doing the same thing, and in the same temporal direction: namely, exponentially accelerating the local expansion of spacetime around them in inverse proportion to the square of mass and distance (viz. the Inverse-Square Law). The difference is that the Big Bang is simply exponentially larger (in terms of total mass-energy) than other supermassive black holes, as well as the most distant such object in our observable past. To see how our reinterpretation of relativity explains exponential expansion of the Universe following the Big Bang as a gravitational field, consider a standard depiction of a gravitational potential (Fig 25).

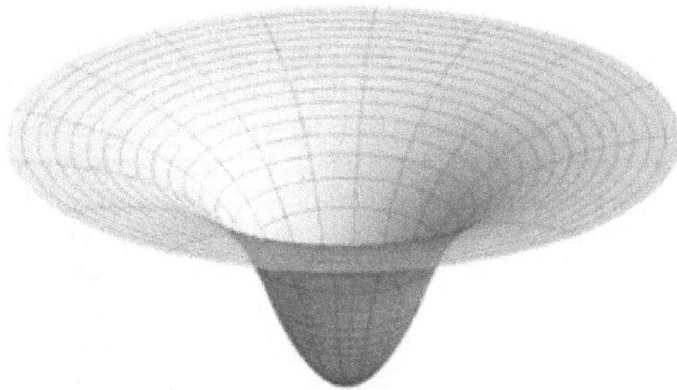

Figure 25: Ordinary Gravitational Potential (fn. 14)

Figure 26 in turn represents the gravitational potentials of the Sun, a neutron star, and wormhole, offset against a representation of "cosmic inflation" turned on its side.

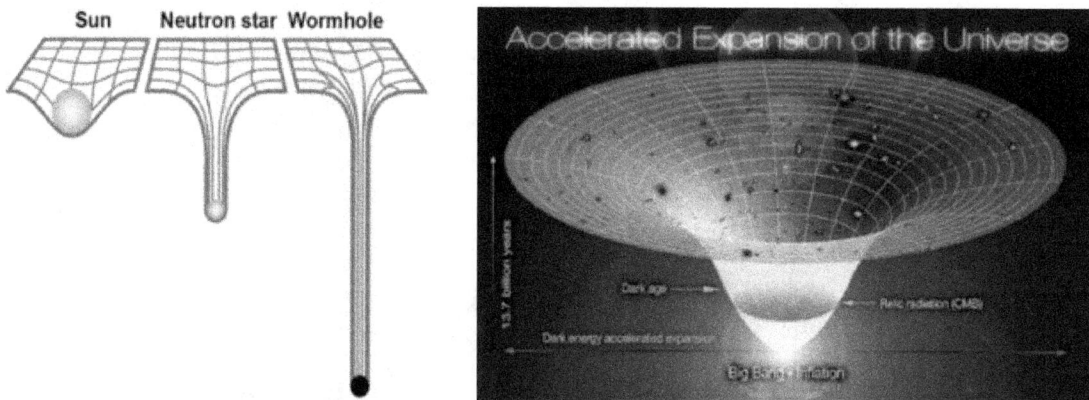

Figure 26: The Big Bang as a Gravity Well (fn. 15)

If our reconceptualization of relativity is correct, then the Big Bang is not a "time-inverted" black hole: it is simply the most massive black hole observable in our light-cone's past, one with such immense mass-energy that it accelerated the expansion of all of the Universe's spacetime near its spacetime horizon (viz. the 'inflationary epoch' in the early Universe) before these effects rapidly dropped off as the rest of the Universe became further removed from the Big Bang singularity (viz. the Inverse-Square Law). On our reinterpretation of the field equations, the Sun, a neutron star, black holes, entire galaxies, and the Big Bang are all (i) doing exactly the same thing, (ii) in the same direction of time: namely, (iii) accelerating the local expansion of dynamic space-time in proportion to their respective amounts of mass-energy (viz. the Inverse Square Law), where (iv) these gravitational effects drop off exponentially in inverse proportion to the square of mass and distance, (v) generating the logarithmic shaped velocity of the Universe's expansion rate extrapolated from observations in Big Bang cosmology.

This implication is in turn supported by another curious relationship between the Big

Bang and ordinary black holes. In 1974, Hawking famously calculated that black holes should emit thermal "blackbody" radiation [34], such that the theorized wavelength of blackbody radiation is asymptotic and inversely proportional to a black hole's mass, leading smaller black holes to emit higher temperatures than larger black holes, which should evaporate more slowly at a lower temperature. Importantly, Hawking radiation is thought to be too small to be directly observable in actual gravitational systems—so empirical confirmation of its existence is thought to remain unsettled today, with the closest instances to confirmation involving experimental analogues of black holes, such as 'sonic black holes' [3] and an optical analogue using lasers [4]. Yet, indirect observations of the Big Bang—via its remnant radiation, the cosmic microwave background—reveal the Big Bang's thermal radiation to have the same type of asymptotic emission curve as Hawking radiation [68] In fact, the fit between the cosmic microwave background and hypothesized emission of Hawking radiation is basically perfect [75] - see Figure 27.

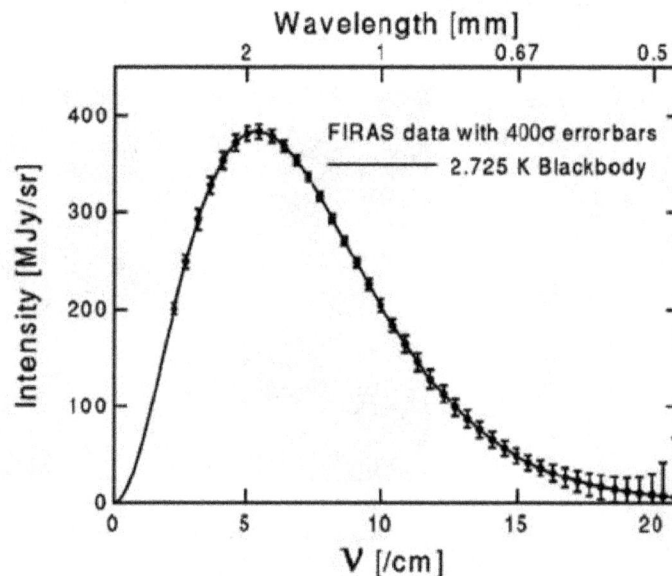

Figure 27: Fit of Cosmic Microwave Background to Theorized Hawking Radiation Curve [73]

If our reconceptualization of relativity is correct, then this is readily explainable: the cosmic microwave background just is Hawking radiation. The Big Bang is not a time-reversed black hole (or white hole). It is simply a black hole in our past light-cone that is so massive that—following Hawking's predictions that larger black holes should dissipate more slowly, inversely proportional to their mass—its energy dissipation has occurred so slowly that it is still observable 13.8 billion years later.

Finally, our interpretation of relativity can explain away yet another curious anomaly. Croker and Weiner [17] argue that when an error in applying general relativity to cosmology is corrected, black holes can be understood as surrounded by a thin halo of dark energy expanding spacetime near the black hole's boundary. This paper's reinterpretation of relatively explains this otherwise baffling result straightforwardly: black holes are not surrounded by a thin crust of (inflationary) "dark energy." Rather, what Croker and Weiner are interpreting as "dark energy" is simply the second-order dynamic expansion of spacetime that our reinterpretation of the field equations hold constitute gravity, as it precisely that kind of accelerated expansion of dynamic space that produces gravitational effects as a measurement artifact.

At this point, we conclude with a rhetorical question. Which of the following two possibilities is more likely, given the history of scientific inquiry?

1. The status-quo hypothesis, which holds, based on the traditional interpretation of general relativity's field equations, that:

  a. The Universe is comprised by a variety of exotic substances (dark matter, dark energy, an inflation field, etc.) that—like the aether, phlogistion, and élan vital—have never been directly observed in any experiment.
  b. The amount and properties of these exotic substances have changed dramatically over the course of the Universe's history for some yet-to-be-understood reason.
  c. In ways that generate a wide variety of anomalies that conflict with the ΛCDM model.
  d. But, despite (a)-(c), our understanding of relativity and the ΛCDM model are correct.

Or,

2. The reinterpretation of general relativity's field equations defended in this paper—which holds that we may explain away all of these cosmological anomalies simply by reinterpreting a central term in the field equations, Λ, as expressing the fundamental nature of gravity as accelerating the local expansion of a second-order spacetime fabric around objects located in an absolute first-order Newtonian space.

Conclusion

This paper's reinterpretation of relativity's physical significance may be misguided. We may have also made mistakes of detail in presenting the interpretation and its various implications. Nevertheless, we believe we have seen ample conceptual reasons to believe there may be something to it. Physics, again, is in crisis. The ΛCDM model of the cosmos—based on the traditional interpretation of the field equations—is rife with theoretical, explanatory, and predictive problems. Dark energy and dark matter, two central elements of the model, are not only astonishingly strange-supposedly constituting nearly all of the Universe, and changing in proportion from one cosmological moment to the next; every experimental search for them to date has yielded null results. Further, a third new theoretical entity widely invoked in order to explain the Universe's exponential inflation just after the Big Bang—a so-called inflation field—multiplies theoretical entities even further, despite the fact that no inflation-field or particle candidates to explain it have ever been experimentally detected.

The alternative interpretation of the field equations we have laid out does away with dark energy, dark matter, and a primordial inflation field, explaining all of the above phenomena in terms of gravity, and gravity in terms of a new interpretation of "Λ": it being the fundamental interaction that mass-energy has on locally accelerating spacetime expansion exponentially across time. We have seen that this new interpretation of relativity holds that gravity does not involve the literal non-Euclidean curvature of spacetime, but instead an accelerated expansion of Euclidean spacetime in a manner that gives rise to observations of "spacetime curvature" (viz. the bending of light, relativistic time and space dilation, etc.) as a measurement artifact generated by the accelerated expansion of a second-order, dynamic Euclidean spacetime fabric against an absolute, first-order Euclidean background. This new interpretation of the field equations may turn out to be incorrect. But, given all of the problems it appears it may be capable of resolving, we submit that the conceptual arguments provided for it warrant further investigation using the specialized methods of mathematical physics.

References

[1] Abi, B., Albahri, T., Al-Kilani, S., Allspach, D., Alonzi, L.P., Anastasi, A., ... & Lusiani, A. (2021). Measurement of the positive muon anomalous magnetic moment to 0.46 ppm. Physical Review Letters, 126(14), 141801-11.

[2] Allanach, B. (2018). Going Nowhere Fast. Aeon, https://aeon.co/essays/has-the-quest-for-top-down-unification-of-physics-stalled.

[3] Barceló, Carlos; Liberati, Stefano; Visser, Matt (2003). "Towards the observation of Hawking radiation in Bose-Einstein condensates". International Journal of Modern Physics A. 18(21): 3735-3745.

[4] Belgiorno, Francesco D.; Cacciatori, Sergio Luigi; Clerici, Matteo; Gorini, Vittorio; Ortenzi, Giovanni; Rizzi, Luca; Rubino, Eleonora; Sala, Vera Giulia; Faccio, Daniele (2010). Hawking radiation from ultrashort laser pulse filaments. Physical Review Letters. 105(20): 203901.

[5] Bertone, G., Hooper, D., & Silk, J. (2005). Particle dark matter: Evidence, candidates and constraints. Physics Reports. 405 (5-6): 279-390.

[6] Binny, J. & Merrifield, M. (1998). Galactic Astronomy. Princeton: Princeton University Press.

[7] Blum, T., Denig, A., Logashenko, I., de Rafael. E., Roberts, B.L., Teubner, T., & Venanzoni, G. (2013). The Muon (g-2) Theory Value: Present and Future. arXiv:1311.2198

[8] Bogdán, Á. & Goulding, A.D. (2015). Connecting dark matter halos with the galaxy center and the supermassive black hole. The Astrophysical Journal, 800(2): 124-34.

[9] Boran, S., Desai, S, Kahya, E. & Woodard, R. (2017). GW170817 Falsifies Dark Matter Emulators. arXiv:1710.06168 [10] Boylan-Kolchin, M. (2022). Stress Testing Λ CDM with High-redshift Galaxy Candidates. arXiv preprint arXiv:2208.01611.

[11] Caldwell, R.R. (2002). A phantom menace? Cosmological consequences of a dark energy component with super-negative equation of state. Physics Letters B. 545(1-2), 23-9.

[12] Callaway, D.J.E. (1988). Triviality Pursuit: Can Elementary Scalar Particles Exist?.

Physics Reports 167(5), 241-320.

[13] Capdevila, B., Crivellin, A., Descotes-Genon, S., Matias, J., & Virto, J. (2018). Patterns of New Physics in b → sℓ + ℓ - transitions in the light of recent data. Journal of High Energy Physics 1(93), 1-21.

[14] CERN (2010). Particle chameleon caught in the act of changing. https://home.cern/news/press-release/cern/particle-chameleon-caught-act-changing.

[15] Cooper, K. (2016). Correlation between galaxy rotation and visible matter puzzles astronomers. Physics World, https://physicsworld.com/a/correlation-between-galaxy-rotation-and-visible-matter-puzzlesastronomers/.

[16] Corbelli E. & Salucci, P. (2000). The extended rotation curve and the dark matter halo of M33. Monthly Notices of the Royal Astronomical Society 311(2), 441-7.

[17] Croker, K.S. & Weiner, J.L. (2019). Implications of Symmetry and Pressure in Friedmann Cosmology. I. Formalism. The Astrophysical Journal, 882(1), 19.

[18] Croswell, K. (2015). The Milky Way's Missing Mass: Partially Found. Scientific American, https://www.scientificamerican.com/article/the-milky-way-s-missing-mass-partially-found/.

[19] Drees, M. & Gerbier, G. (2015). Dark Matter. Chin. Phys. C. 38: 090001.

[20] Dymnikova, I. (2019). Universes Inside a Black Hole with the de Sitter Interior. Universe, 5(5), 1-11.

[21] Einstein, A. (1940). The Fundaments of Theoretical Physics, Science, 91(2369), 487-92.

[22] Einstein, A. [1922b]. The Meaning of Relativity. London: Routledge, 2003.

[23] Einstein, A. (1917). Cosmological considerations on the general theory of relativity. Sitzungsberichte der Königlich Preußischen Akademie der Wissenschaften. Berlin, DE. part 1, 142-152.

[24] Einstein, A. (1915). The field equations of gravitation. Sitzungsberichte der Preussischen Akademie der Wissenschaften zu Berlin, 844-7.

[25] European Southern Observatory (2017). Science Release: Dark Matter Less Influential in Galaxies in Early Universe. https://www.eso.org/public/news/eso1709/.

[26] Faber, S.M. & Jackson, R.E. (1976). Velocity dispersions and mass-to-light ratios for elliptical galaxies. The Astrophysical Journal 204, 668-83.

[27] Francis, M. (2013). First Planck Results: The Universe is Still Weird and Interesting. Ars Technica, https://arstechnica.com/science/2013/03/first-planck-results-the-universe-is-still-weird-andinteresting/.

[28] Gentile, G. & Salucci, P. (2004). The cored distribution of dark matter in spiral galaxies. Monthly Notices of the Royal Astronomical Society 351, 903-22.

[29] Genzel, R., Schreiber, N.F., Übler, H., Lang, P., Naab, T., Bender, R., ... & Beifiori, A. (2017). Strongly baryondominated disk galaxies at the peak of galaxy formation ten billion years ago. Nature 543(7645), 397401.

[30] Ghez, A.M., Salim, S., Weinberg, N.N., Lu, J.R., Do, T., Dunn, J.K., Matthews, K. ... Becklin, E.E. (2008). Measuring Distance and Properties of the Milky Way's Central Supermassive Black Hole with Stellar Orbits. Astrophysical Journal. 689(2), 1044-1062.

[31] Gillessen, S., Eisenhauer, F. Trippe, S., Alexander, T., Genzel, R., Martins, F., & Ott, T. (2009). Monitoring stellar orbits around the Massive Black Hole in the Galactic Center. The Astrophysical Journal 692(2), 1075-109.

[32] Guth, A.H. (1997). The Inflationary Universe: The Quest for a New Theory of Cosmic Origins. Basic Books.

[33] Guth, A.H. (1981). Inflationary universe: A possible solution to the horizon and flatness problems. Physical Review D. 23(2), 347-56.

[34] Hawking, S.W. (1974). Black hole explosions?. Nature, 248(5443), 30-31.

[35] Hawking, S.W. (1976). Black Holes and Thermodynamics. Physical Review D. 13(2), 191-7.

[36] Hubble, E. (1929). A Relation between Distance and Radial Velocity among Extra-Galactic Nebulae. Proceedings of the National Academy of Sciences of the United States of America 15(3), 168-73.

[37] International Astronomical Union. (2018). IAU members vote to recommend renaming the Hubble law as the Hubble-Lemaître law,
https://www.iau.org/news/pressreleases/detail/iau1812/.

[38] Kuhn, T. (1962). The Structure of Scientific Revolutions. Chicago: University of Chicago Press.

[39] Laplane L., Mantovani, P., Adolphs, R., Chang, H., Mantovani, A., McFall-Ngai, M., Rovelli, C., Sober, E, and Pradeu, T. (2019). Opinion: Why Science Needs Philosophy. PNAS March 5, 2019 116(10), 3948-52.

[40] Letzter, R. (2019). The Universe is expanding faster than we thought, and no one can explain the 'mismatch.' NBCNews.com, https://www.nbcnews.com/mach/science/universe-expanding-faster-we-thought-no-one-can-explain-mismatch-ncna998901.

[41] López-Corredoira, M., Allende Prieto, C., Garzón, F., Wang, H., Liu, C. & Deng, L. (2018). Disk stars in the Milky Way detected beyond 25 kpc from its center. Astronomy & Astrophysics. 612: L8.

[42] Lu, R., Krichbaum, T.P., Roy, A.L., Fish, V.L., Doeleman, S.S., Johnson, M.D. ... Ziurys, L.M. (2018). Detection of intrinsic source structure at ~ 3 Schwarzschild radii with Millimeter-VLBI observations of Sgr A*. Astrophysical Journal 859(1), 60-71.

[43] Lykken, J. & Spiropulu (2014). Supersymmetry and the Crisis in Physics. Scientific American, https://www.scientificamerican.com/article/supersymmetry-and-the-crisis-in-physics/.

[44] Mandelbaum, R.F. (2019). Hubble Measurements Confirm There's Something Weird About How the Universe Is Expanding. Gizmodo. https://gizmodo.com/hubble-measurements-confirm-theres-something-weird-abou-1834339830.

[45] McGaugh, S.S., Lelli, F., & Schombert, J.M. (2016). Radial acceleration relation in rotationally supported galaxies. Physical Review Letters 117(20), 201101.

[46] Milgrom, M. (1983). A modification of the Newtonian dynamics as a possible alternative to the hidden mass hypothesis. Astrophysical Journal 270, 365-70.

[47] Misner, C.W., Thorne, KS.; Wheeler, J.A. & Archibald, J. (1973). Gravitation. San Francisco: W.H. Freeman.

[48] Mitrica, D. (2017). Surprising Link Found Between Dark Matter and Black Holes. ZME Science, https://www.zmescience.com/space/observations/dark-matter-black-hole-20022015/.

[49] Moskowitz, C. (2010). Einstein's 'Biggest Blunder' Turns Out to Be Right. Space.com, https://www.space.com/9593-einstein-biggest-blunder-turns.html.

[50] NASA (2019). Black Hole Image Makes History; NASA Telescopes Coordinated Obser-

vations.
https://www.nasa.gov/mission_pages/chandra/news/black-hole-image-makes-history.

[51] NASA (2015). What is the Ultimate Fate of the Universe?,
https://map.gsfc.nasa.gov/universe/uni_fate.html.

[52] Peacock, J.A., Cole, S., Norberg, P., Baugh, C.M., Bland-Hawthorne, J., Bridged, T. Taylor, K. (2001). A measurement of the cosmological mass density from clustering in the 2dF Galaxy Redshift Survey. Nature. 410 (6825), 169-73.

[53] Peebles, P.J.E. & Ratra, B. (2003). The cosmological constant and dark energy. Reviews of Modern Physics 75(2), 559-606.

[54] Percy, J. (1993). Black Holes. Astronomical Society of the Pacific,
https://astrosociety.org/file_download/inline/a2ea7181-f4d5-4066-8490-3a2e7c44f28f.

[55] Persic, M. & Salucci, P. (1992). The baryon content of the Universe. Monthly Notices of the Royal Astronomical Society 258(1), 14P-18P.

[56] Pohl, R., Gilman, R., Miller, G.A., & Pachucki, K. (2013). Muonic hydrogen and the proton radius puzzle. Annu. Rev. Nucl. Part. Sci. 63, 175-204.

[57] Pössel, M. (2005). The elevator, the rocket, and gravity: the equivalence principle. Einstein Online Vol. 01 (2005): 1009,
http://www.einstein-online.info/spotlights/equivalence_principle.html.

[58] Powell, C.S. (2019). The universe may be a billion years younger than we thought. Scientists are scrambling to figure out why. NBCNews.com, https://www.nbcnews.com/mach/science/universe-may-be-billion-years-younger-we-thought-scientists-are-ncna1005541.

[59] Quine, W.V.O. (2008). Chapter 31: Three indeterminacies. Confessions of a Confirmed Extensionalist: And Other Essays. Harvard University Press, 368-386.

[60] Ratra, P. & Peebles, L. (1988). Cosmological consequences of a rolling homogeneous scalar field. Physical Review D. 37(12), 3406.

[61] Refregier, A. (2003). Weak gravitational lensing by large-scale structure. Annual Review of Astronomy and Astrophysics. 41(1), 645-68.

[62] Riess, A.G., Casertano, S., Yuan, W., Macri, L.M., & Scolnic, D. (2019). Large Magellanic Cloud Cepheid Standards Provide a 1% Foundation for the Determination of the Hubble Constant and Stronger Evidence for Physics Beyond LambdaCDM. arXiv preprint arXiv:1903.07603.

[63] Riess, A., Filippenko, A.V., Challis, P., Clocchiatti, A., Diercks, A., Garnavich, P.M. ... Tonry, J. (1998). Observational Evidence from Supernovae for an Accelerating Universe and a Cosmological Constant. The Astronomical Journal 116(3), 1009-38.

[64] Riess, A. G., Yuan, W., Macri, L. M., Scolnic, D., Brout, D., Casertano, S., ... & Zheng, W. (2022). A comprehensive measurement of the local value of the hubble constant with 1 km/s/Mpc uncertainty from the hubble space telescope and the SHOES team. arXiv preprint arXiv:2112.04510v2

[65] Rovelli, C. (2018). Physics needs philosophy. Philosophy needs physics. Foundations of Physics 48.5, 481491.

[66] Salvio, S. (2014). Agravity. JHEP. 6(6): 080. arXiv:1403.4226

[67] Stanford, K. (2017). Underdetermination of Scientific Theory. In E.N. Zalta (ed.), The Stanford Encyclopedia of Philosophy (Winter 2017 Edition), Edward N. Zalta (ed.), https://plato.stanford.edu/archives/win2017/entries/scientific-underdetermination/.

[68] Strobel, N. (2019). Cosmic Microwave Background Radiation. https://www.astronomynotes.com/cosmolgy/s5.htm

[69] Sushkov, A.O., Kim, W.J., Dalvit, D A.R. & Lamoreaux, S.K. (2011). New Experimental Limits on NonNewtonian Forces in the Micrometer Range. Physical Review Letters 107(17), 171101.

[70] Taylor, A.N., Broadhurst T.J., Benítez, N. & van Kampen E. (1998). Gravitational Lens Magnification and the Mass of Abell 1689. The Astrophysical Journal 501(2), 539-53.

[71] Trimble, V. (1987). Existence and nature of dark matter in the universe. Annual Review of Astronomy and Astrophysics 25, 425-472.

[72] Wall, M. (2019). The Universe Is Expanding So Fast We Might Need New Physics to Explain It. Space.com. https://www.space.com/universe-expanding-fast-new-physics.html.

[73] Wright, K. (2018). Synopsis: revised prediction for Mercury's orbit. American Physical Society. https://physics.aps.org/synopsis-for/10.1103/PhysRevLett.120.191101.

[74] Wu, X., Chiueh, T., Fang, L. & Xue, Y. (1998). A comparison of different cluster mass estimates: consistency or discrepancy?. Monthly Notices of the Royal Astronomical Society 301(3), 861-7.

[75] Wuensche, C. A., & Villela, T. (2010). 25 years of Cosmic Microwave Background research at INPE. arXiv preprint arXiv:1002.4902.

Notes

[1] File:Noneuclid.svg by Joshuabowman is licensed under the Creative Commons Attribution-Share Alike 3.0
Unported license, https://commons.wikimedia.org/wiki/File:Noneuclid.svg.

[2] File:Cartesian-Plane.png is licensed under the Creative Commons Attribution Share Alike License, https://wikieducator.org/File:Cartesian-Plane.png.

[3] File:Inverse square law.svg by Borb is licensed under the GNU Free Document License, https://commons.wikimedia.org/wiki/File:Inverse_square_law.svg.

[4] File:Inverse proportionality function plot.gif is licensed under Creative Commons CC0 1.0 Universal Public Domain Dedication. https://commons.wikimedia.org/wiki/File:Inverse_proportionality function_plot.gif.

[5] File:Exp.svg by Peter John Acklam is licensed under the GNU Free Document License. https://commons.wikimedia.org/wiki/File:Exp.svg.

[6] File:Colors and wavelengths.png by Sharon Bewick is is licensed under the Creative Commons Attribution-Share Alike 3.0 Unported license.
https://commons.wikimedia.org/wiki/File:Colors_and_wavelengths.png. Alterations were made by this paper's author, separating image into distinct images.

[7] File:Basic human drawing.png by Wikifanboy000 is is licensed under the Creative Commons Attribution-Share Alike 4.0 International license. https://commons.wikimedia.org/wiki/File:Basic human drawing.png.

[8] File:Hubble constant.JPG by Brews ohare is licensed under the Creative Commons Attribution-Share Alike 3.0 Unported license.

[9] File:Galaxy rotation under the influence of dark matter.ogv By Ingo Berg is licensed under

5 Variational Formulation of Non-Barotropic Relativistic Fluid Dynamics and the Canonical Energy-Momentum Tensor of Flows

Asher Yahalom

Abstract   We describe a variational formulation of non-barotropic relativistic fluid dynamics and derive the canonical energy-momentum tensor of non-barotropic flows, some cosmological implications are discussed.

Keywords: Fluid dynamics, Non-Barotropic Flows, Variational Analysis, Canonical Energy Momentum Tensor

1   Introduction

A comprehensive introduction to the subject of the variational formalism of non-relativistic fluid dynamics and quantum mechanics and their deep interconnections is given in [1, 2] and will not be repeated here.

The original work of Clebsch [3, 4] and all the following publications assume a non-relativistic fluids in which the velocity of the flow is much slower than the speed of light in vacuum c. This is of course to be expected as the work of Clebsch preceded Einstein's work on special relativity by forty eight years. Practically, relativistic flows are hardly encountered on earth.

The standard approach to relativistic flows is based on the energy-momentum tensor [7, 8, 9], however, this approach is not rigorous because the definition of an energy-momentum tensor can only be done if a Lagrangian density is provided [10]. However, no Lagrangian density was known for relativistic flows. In a recent work I have expanded Clebsch work to relativistic flows [5] and thus amended this lacuna with a derived Lagrangian density for a relativistic flow from which one can obtain rigorously the energy-momentum tensor of high velocity flows. A connection between Relativistic flows and the Dirac theory of the electron through the concept of Fisher information is elucidated in [6]. However, previous work did not consider non-barotropic Eulerian flows as the emphasis was on making the connection with quantum mechanics, this is amended in the current work.

We will not repeat the formulation of a variational principle for a relativistic charged classical particle with a vector potential interaction and a system of charged particles given in [5]. However, we shall use the notation of [5] whenever it is needed.

We shall start with the Eckart [11] Lagrangian variational principles generalized for a relativistic charged fluid. This will be followed by the general theory of the canonical energy-momentum tensor. We then derive the energy momentum tensor related to the

Kyley Ewing (Ed.), *Spacetime Conference 2024. Selected peer-reviewed papers presented at the Seventh International Conference on the Nature and Ontology of Spacetime, 16 - 19 September 2024, Albena, Bulgaria* (Minkowski Institute Press, Montreal 2025). ISBN 978-1-998902-44-6 (softcover), ISBN 978-1-998902-45-3 (ebook).

Eckart Lagrangian. Next we introduce an Eulerian-Clebsch variational principle for a non-barotropic relativistic charged fluid which is followed by derivation of the energy-momentum tensor of the same. Finally we discuss the cosmological implications of the energy-momentum tensor.

2 The Lagrangian Description of a Relativistic Charged Fluid

2.1 Action and Lagrangian

Dynamics of a flow are specified by its composition and the forces which act on it. The fluid is composed of "fluid elements" [11, 12]. A "fluid element" is a point particle with an infinitesimal mass $dM_{\vec{\alpha}}$, infinitesimal charge $dQ_{\vec{\alpha}}$, position four vector $x_{\vec{\alpha}\nu}(\tau_{\vec{\alpha}})$ and $u_{\vec{\alpha}\nu}(\tau_{\vec{\alpha}}) \equiv \frac{dx_{\vec{\alpha}\nu}(\tau_{\vec{\alpha}})}{d\tau_{\vec{\alpha}}}$. Here the continuous vector label $\vec{\alpha}$ comes instead of the discrete index n used in the discrete case (see section 2 of [5]). However, the "fluid element" is not a proper point particle since it has an infinitesimal volume $dV_{\vec{\alpha}}$, infinitesimal entropy $dS_{\vec{\alpha}}$, and an infinitesimal internal energy $dE_{in\ \vec{\alpha}}$. The action for each "fluid element" is following equation (1) of [5] is of the form:

$$
dA_{\vec{\alpha}} = -dM_{\vec{\alpha}}c \int d\tau_{\vec{\alpha}} - dQ_{\vec{\alpha}} \int A^\mu(x_{\vec{\alpha}}^\nu)dx_{\mu\vec{\alpha}} + dA_{in\ \vec{\alpha}},
$$

$$
dA_{in\ \vec{\alpha}} \equiv -\int dE_{in\ \vec{\alpha}}dt. \tag{1}
$$

The Lagrangian for each "fluid element" can be derived from the above expression as follows:

$$
dA_{\vec{\alpha}} = \int_{t1}^{t2} dL_{\vec{\alpha}}dt, \qquad dL_{\vec{\alpha}} \equiv dL_{k\vec{\alpha}} + dL_{i\vec{\alpha}} - dE_{in\ \vec{\alpha}}
$$

$$
dL_{k\vec{\alpha}} \equiv -\frac{dM_{\vec{\alpha}}c^2}{\gamma_{\vec{\alpha}}} \simeq \frac{1}{2}dM_{\vec{\alpha}}\ v_{\vec{\alpha}}(t)^2 - dM_{\vec{\alpha}}c^2
$$

$$
dL_{i\vec{\alpha}} \equiv dQ_{\vec{\alpha}}\left(\vec{A}(\vec{x}_{\vec{\alpha}}(t),t)\cdot\vec{v}_{\vec{\alpha}}(t) - \phi(\vec{x}_{\vec{\alpha}}(t),t)\right). \tag{2}
$$

The action and Lagrangian of the entire fluid, is integrated over all possible $\vec{\alpha}$'s:

$$
L = \int_{\vec{\alpha}} dL_{\vec{\alpha}}
$$

$$
A = \int_{\vec{\alpha}} dA_{\vec{\alpha}} = \int_{\vec{\alpha}}\int_{t1}^{t2} dL_{\vec{\alpha}}dt = \int_{t1}^{t2}\int_{\vec{\alpha}} dL_{\vec{\alpha}}dt = \int_{t1}^{t2} Ldt. \tag{3}
$$

We define a density by dividing a fluid element quantity by its volume. This is done for the Lagrangian, mass, charge and internal energy of every fluid element by introducing the following symbols:

$$
\mathcal{L}_{\vec{\alpha}} \equiv \frac{dL_{\vec{\alpha}}}{dV_{\vec{\alpha}}}, \quad \rho_{\vec{\alpha}} \equiv \frac{dM_{\vec{\alpha}}}{dV_{\vec{\alpha}}}, \quad \rho_{c\vec{\alpha}} \equiv \frac{dQ_{\vec{\alpha}}}{dV_{\vec{\alpha}}}, \quad e_{in\ \vec{\alpha}} \equiv \frac{dE_{in\ \vec{\alpha}}}{dV_{\vec{\alpha}}} \tag{4}
$$

Every quantity of the density type is a function of $\vec{x}$, in which the "fluid element" labelled $\vec{\alpha}$ is located in time t, for example:

$$\rho(\vec{x}, t) \equiv \rho(\vec{x}_{\vec{\alpha}}(t), t) \equiv \rho_{\vec{\alpha}}(t). \tag{5}$$

We also define "specific" quantities by dividing attribute of the "fluid element" by its mass, for example a specific internal energy $\varepsilon_{\vec{\alpha}}$ is:

$$\varepsilon_{\vec{\alpha}} \equiv \frac{dE_{in\ \vec{\alpha}}}{dM_{\vec{\alpha}}} \quad \Rightarrow \quad \rho_{\vec{\alpha}}\varepsilon_{\vec{\alpha}} = \frac{dM_{\vec{\alpha}}}{dV_{\vec{\alpha}}} \frac{dE_{in\ \vec{\alpha}}}{dM_{\vec{\alpha}}} = \frac{dE_{in\ \vec{\alpha}}}{dV_{\vec{\alpha}}} = e_{in\ \vec{\alpha}} \tag{6}$$

Thus we can partition the Lagrangian density as follows:

$$
\begin{aligned}
\mathcal{L}_{\vec{\alpha}} &= \frac{dL_{\vec{\alpha}}}{dV_{\vec{\alpha}}} = \frac{dL_{k\vec{\alpha}}}{dV_{\vec{\alpha}}} + \frac{dL_{i\vec{\alpha}}}{dV_{\vec{\alpha}}} - \frac{dE_{in\ \vec{\alpha}}}{dV_{\vec{\alpha}}} = \mathcal{L}_{k\vec{\alpha}} + \mathcal{L}_{i\vec{\alpha}} - e_{in\ \vec{\alpha}} \\
\mathcal{L}_{k\vec{\alpha}} &\equiv -\frac{\rho_{\vec{\alpha}}c^2}{\gamma_{\vec{\alpha}}} \simeq \frac{1}{2}\rho_{\vec{\alpha}}v_{\vec{\alpha}}(t)^2 - \rho_{\vec{\alpha}}c^2, \\
\mathcal{L}_{i\vec{\alpha}} &\equiv \rho_{c\vec{\alpha}}\left(\vec{A}(\vec{x}_{\vec{\alpha}}(t), t) \cdot \vec{v}_{\vec{\alpha}}(t) - \varphi(\vec{x}_{\vec{\alpha}}(t), t)\right).
\end{aligned}
\tag{7}
$$

We can thus write the fluid Lagrangian as a spatial integral:

$$L = \int_{\vec{\alpha}} dL_{\vec{\alpha}} = \int_{\vec{\alpha}} \mathcal{L}_{\vec{\alpha}} dV_{\vec{\alpha}} = \int \mathcal{L}(\vec{x}, t)d^3x \tag{8}$$

which will be used in later section concerned with the Eulerian representation of the fluid.

2.2 Variational Analysis

Let us introduce the symbols $\Delta\vec{x}_{\vec{\alpha}} \equiv \vec{\xi}_{\vec{\alpha}}$ to denote a variation of the trajectory $\vec{x}_{\vec{\alpha}}(t)$. Thus:

$$\Delta\vec{v}_{\vec{\alpha}}(t) = \Delta\frac{d\vec{x}_{\vec{\alpha}}(t)}{dt} = \frac{d\Delta\vec{x}_{\vec{\alpha}}(t)}{dt} = \frac{d\vec{\xi}_{\vec{\alpha}}(t)}{dt}. \tag{9}$$

And thus according to equation (9) of [5]:

$$\Delta\left(\frac{1}{\gamma_{\vec{\alpha}}}\right) = -\frac{\gamma_{\vec{\alpha}}\vec{v}_{\vec{\alpha}}(t)}{c^2}\frac{d\vec{\xi}_{\vec{\alpha}}(t)}{dt}, \qquad \Delta\gamma_{\vec{\alpha}} = \frac{\gamma_{\vec{\alpha}}^3\vec{v}_{\vec{\alpha}}(t)}{c^2}\frac{d\vec{\xi}_{\vec{\alpha}}(t)}{dt}. \tag{10}$$

An "ideal fluid" is defined by the fact that the "fluid element" does exchange mass, nor electric charge, nor heat with other elements, or in a variational form:

$$\Delta dM_{\vec{\alpha}} = \Delta dQ_{\vec{\alpha}} = \Delta dS_{\vec{\alpha}} = 0. \tag{11}$$

According to thermodynamics a change in the internal energy of a "fluid element" satisfies the equation below in the particle's rest frame:

$$\Delta dE_{in\ \vec{\alpha}0} = T_{\vec{\alpha}0}\Delta dS_{\vec{\alpha}0} - P_{\vec{\alpha}0}\Delta dV_{\vec{\alpha}0}. \tag{12}$$

In the above the first term describes the heating of the "fluid element" while the second term is a manifestation the work done by the "fluid element" on neighbouring elements. $T_{\vec{\alpha}0}$ denotes the temperature of the "fluid element" in the rest frame, and $P_{\vec{\alpha}0}$ is the pressure of the same. As the rest mass of the fluid element does not change and does not depend on any specific frame we may divide the above expression by $dM_{\vec{\alpha}}$ to derive the variation of the specific energy:

$$
\begin{aligned}
\Delta \varepsilon_{\vec{\alpha}0} &= \Delta \frac{dE_{in\ \vec{\alpha}0}}{dM_{\vec{\alpha}}} = T_{\vec{\alpha}0} \Delta \frac{dS_{\vec{\alpha}0}}{dM_{\vec{\alpha}}} - P_{\vec{\alpha}0} \Delta \frac{dV_{\vec{\alpha}0}}{dM_{\vec{\alpha}}} \\
&= T_{\vec{\alpha}0} \Delta s_{\vec{\alpha}0} - P_{\vec{\alpha}0} \Delta \frac{1}{\rho_{\vec{\alpha}0}} = T_{\vec{\alpha}0} \Delta s_{\vec{\alpha}0} + \frac{P_{\vec{\alpha}0}}{\rho_{\vec{\alpha}0}^2} \Delta \rho_{\vec{\alpha}0}. \quad s_{\vec{\alpha}0} \equiv \frac{dS_{\vec{\alpha}0}}{dM_{\vec{\alpha}}}
\end{aligned}
\tag{13}
$$

$s_{\vec{\alpha}0}$ is the specific entropy of the fluid element in its rest frame. It follows that (we suppress the indices $\vec{\alpha}$ below):

$$
\frac{\partial \varepsilon_0}{\partial s_0} = T_0, \qquad \frac{\partial \varepsilon_0}{\partial \rho_0} = \frac{P_0}{\rho_0^2}.
\tag{14}
$$

Another useful thermodynamic quantity is the Enthalpy defined for a fluid element in its rest frame as:

$$
dW_{\vec{\alpha}0} = dE_{in\ \vec{\alpha}0} + P_{\vec{\alpha}0} dV_{\vec{\alpha}0}.
\tag{15}
$$

and the specific enthalpy:

$$
w_{\vec{\alpha}0} = \frac{dW_{\vec{\alpha}0}}{dM_{\vec{\alpha}}} = \frac{dE_{in\ \vec{\alpha}0}}{dM_{\vec{\alpha}}} + P_{\vec{\alpha}0} \frac{dV_{\vec{\alpha}0}}{dM_{\vec{\alpha}}} = \varepsilon_{\vec{\alpha}0} + \frac{P_{\vec{\alpha}0}}{\rho_{\vec{\alpha}0}}.
\tag{16}
$$

Combining the above with equation (14) we obtain the useful properties:

$$
w_0 = \varepsilon_0 + \frac{P_0}{\rho_0} = \varepsilon_0 + \rho_0 \frac{\partial \varepsilon_0}{\partial \rho_0} = \frac{\partial (\rho_0 \varepsilon_0)}{\partial \rho_0}.
\tag{17}
$$

$$
\begin{aligned}
\frac{\partial w_0}{\partial \rho_0} &= \frac{\partial (\varepsilon_0 + \frac{P_0}{\rho_0})}{\partial \rho_0} = -\frac{P_0}{\rho_0^2} + \frac{1}{\rho_0} \frac{\partial P_0}{\partial \rho_0} + \frac{\partial \varepsilon_0}{\partial \rho_0} = -\frac{P_0}{\rho_0^2} + \frac{1}{\rho_0} \frac{\partial P_0}{\partial \rho_0} + \frac{P_0}{\rho_0^2} \\
&= \frac{1}{\rho_0} \frac{\partial P_0}{\partial \rho_0}.
\end{aligned}
\tag{18}
$$

$$
\vec{\nabla} w_0 = T_0 \vec{\nabla} s_0 + \frac{1}{\rho_0} \vec{\nabla} P_0.
\tag{19}
$$

For an ideal fluid, we neglect heat conduction and radiation, and thus only convection is considered. Thus $\Delta dS_{\vec{\alpha}0} = 0$ and it follows that:

$$
\Delta dE_{in\ \vec{\alpha}0} = -P_0 \Delta dV_{\vec{\alpha}0}.
\tag{20}
$$

Let us now establish some relations between the rest frame and any other frame in which the fluid element is in motion (this frame is sometimes denoted the "laboratory" frame). First we notice that at the rest frame there is no velocity (by definition), hence according to equation (9) of [5]:

$$
d\tau = cdt_0 = cdt\sqrt{1 - \frac{v^2}{c^2}} = \frac{cdt}{\gamma} \quad \Rightarrow \quad dt_0 = \frac{dt}{\gamma}.
\tag{21}
$$

It is well known that the four volume is Lorentz invariant, hence:

$$dV_0 dt_0 = dV dt = dV dt_0 \gamma, \qquad \Rightarrow dV_0 = \gamma dV. \tag{22}$$

Thus:

$$\rho_0 = \frac{dM}{dV_0} = \frac{1}{\gamma}\frac{dM}{dV} = \frac{\rho}{\gamma}, \qquad \Rightarrow \quad \rho = \gamma \rho_0. \tag{23}$$

Moreover, the action given in equation (1) is Lorentz invariant, thus:

$$dE_{in\ \vec{\alpha}0}dt_0 = dE_{in\ \vec{\alpha}}dt = dE_{in\ \vec{\alpha}}dt_0\gamma$$
$$\Rightarrow \quad dE_{in\ \vec{\alpha}0} = \gamma dE_{in\ \vec{\alpha}}, dE_{in\ \vec{\alpha}} = \frac{dE_{in\ \vec{\alpha}0}}{\gamma} \tag{24}$$

We can now vary the internal energy of a fluid element:

$$\Delta dE_{in\ \vec{\alpha}} = \Delta\left(\frac{1}{\gamma}\right)dE_{in\ \vec{\alpha}0} + \frac{1}{\gamma}\Delta dE_{in\ \vec{\alpha}0}. \tag{25}$$

Taking into account equation (20) and equation (22) we obtain:

$$\Delta dE_{in\ \vec{\alpha}} = \Delta\left(\frac{1}{\gamma}\right)dE_{in\ \vec{\alpha}0} - \frac{1}{\gamma}P_0\Delta dV_{\vec{\alpha}0} = \Delta\left(\frac{1}{\gamma}\right)dE_{in\ \vec{\alpha}0} - \frac{1}{\gamma}P_0\Delta(\gamma dV_{\vec{\alpha}}). \tag{26}$$

Thus using the definition of enthalpy given in equation (15) we may write:

$$\Delta dE_{in\ \vec{\alpha}} = \Delta\left(\frac{1}{\gamma}\right)(dE_{in\ \vec{\alpha}0} + P_0 dV_{\vec{\alpha}0}) - P_0\Delta dV_{\vec{\alpha}} = \Delta\left(\frac{1}{\gamma}\right)dW_{\vec{\alpha}0} - P_0\Delta dV_{\vec{\alpha}}. \tag{27}$$

We shall now vary the volume element. At time t the volume of the fluid element is:

$$dV_{\vec{\alpha},t} = d^3x(\vec{\alpha}, t) \tag{28}$$

The Jacobian relates this to the same element at $t = 0$:

$$d^3x(\vec{\alpha}, t) = J d^3x(\vec{\alpha}, 0), \qquad J \equiv \vec{\nabla}_0 x_1 \cdot (\vec{\nabla}_0 x_2 \times \vec{\nabla}_0 x_3) \tag{29}$$

$\vec{\nabla}_0$ is calculated with respect to the $t = 0$ coordinates of the fluid elements:

$$\vec{\nabla}_0 \equiv (\frac{\partial}{\partial x(\vec{\alpha}, 0)_1}, \frac{\partial}{\partial x(\vec{\alpha}, 0)_2}, \frac{\partial}{\partial x(\vec{\alpha}, 0)_3}). \tag{30}$$

An interesting point is that the construction of the volume element of space from 3 vectors, constructing a Jacobian as in equation (29), is very similar to the construction of a four dimensional volume element from 4 scalars and considering also a Jacobian of the transformation of space time coordinates to scalar field space, as done long time ago in [17, 18, 19, 20]. Thus:

$$\Delta dV_{\vec{\alpha},t} = \Delta d^3x(\vec{\alpha}, t) = \Delta J\ d^3x(\vec{\alpha}, 0) = \frac{\Delta J}{J}d^3x(\vec{\alpha}, t) = \frac{\Delta J}{J}dV_{\vec{\alpha},t},$$
$$(\Delta d^3x(\vec{\alpha}, 0) = 0). \tag{31}$$

The variation of J can thus be derived as:

$$\Delta J = \vec{\nabla}_0 \Delta x_1 \cdot (\vec{\nabla}_0 x_2 \times \vec{\nabla}_0 x_3) + \vec{\nabla}_0 x_1 \cdot (\vec{\nabla}_0 \Delta x_2 \times \vec{\nabla}_0 x_3)$$
$$+ \vec{\nabla}_0 x_1 \cdot (\vec{\nabla}_0 x_2 \times \vec{\nabla}_0 \Delta x_3), \tag{32}$$

Now:

$$\vec{\nabla}_0 \Delta x_1 \cdot (\vec{\nabla}_0 x_2 \times \vec{\nabla}_0 x_3) = \vec{\nabla}_0 \xi_1 \cdot (\vec{\nabla}_0 x_2 \times \vec{\nabla}_0 x_3)$$
$$= \partial_k \xi_1 \vec{\nabla}_0 x_k \cdot (\vec{\nabla}_0 x_2 \times \vec{\nabla}_0 x_3) = \partial_1 \xi_1 \vec{\nabla}_0 x_1 \cdot (\vec{\nabla}_0 x_2 \times \vec{\nabla}_0 x_3) = \partial_1 \xi_1 J.$$
$$\vec{\nabla}_0 x_1 \cdot (\vec{\nabla}_0 \Delta x_2 \times \vec{\nabla}_0 x_3) = \vec{\nabla}_0 x_1 \cdot (\vec{\nabla}_0 \xi_2 \times \vec{\nabla}_0 x_3)$$
$$= \partial_k \xi_2 \vec{\nabla}_0 x_1 \cdot (\vec{\nabla}_0 x_k \times \vec{\nabla}_0 x_3) = \partial_2 \xi_2 \vec{\nabla}_0 x_1 \cdot (\vec{\nabla}_0 x_2 \times \vec{\nabla}_0 x_3) = \partial_2 \xi_2 J.$$
$$\vec{\nabla}_0 x_1 \cdot (\vec{\nabla}_0 x_2 \times \vec{\nabla}_0 \Delta x_3) = \vec{\nabla}_0 x_1 \cdot (\vec{\nabla}_0 x_2 \times \vec{\nabla}_0 \xi_3)$$
$$= \partial_k \xi_3 \vec{\nabla}_0 x_1 \cdot (\vec{\nabla}_0 x_2 \times \vec{\nabla}_0 x_k) = \partial_3 \xi_3 \vec{\nabla}_0 x_1 \cdot (\vec{\nabla}_0 x_2 \times \vec{\nabla}_0 x_3) = \partial_3 \xi_3 J. \tag{33}$$

Thus:

$$\Delta J = \partial_1 \xi_1 J + \partial_2 \xi_2 J + \partial_3 \xi_3 J = \vec{\nabla} \cdot \vec{\xi} \, J, \qquad \Delta dV_{\vec{\alpha},t} = \vec{\nabla} \cdot \vec{\xi} \, dV_{\vec{\alpha},t}. \tag{34}$$

So the variation of the internal energy of equation (27) can be written as:

$$\Delta dE_{in\ \vec{\alpha}} = \Delta \left(\frac{1}{\gamma} \right) dW_{\vec{\alpha}0} - P_0 \vec{\nabla} \cdot \vec{\xi} \, dV_{\vec{\alpha},t}. \tag{35}$$

Taking into account equation (10) this takes the form:

$$\Delta dE_{in\ \vec{\alpha}} = -P_{\vec{\alpha}0} \vec{\nabla} \cdot \vec{\xi}_{\vec{\alpha}} \, dV_{\vec{\alpha},t} - \frac{\gamma_{\vec{\alpha}} \vec{v}_{\vec{\alpha}}(t)}{c^2} dW_{\vec{\alpha}0} \cdot \frac{d\vec{\xi}_{\vec{\alpha}}(t)}{dt}. \tag{36}$$

The variation of internal energy is the only new calculation with respect to the calculation done for a system of particles described previously, thus the rest of the analysis is trivial. Varying equation (1) we thus obtain:

$$\Delta d\mathcal{A}_{\vec{\alpha}} = \int_{t1}^{t2} \Delta dL_{\vec{\alpha}} dt, \qquad \Delta dL_{\vec{\alpha}} = \Delta dL_{k\vec{\alpha}} + \Delta dL_{i\vec{\alpha}} - \Delta dE_{in\ \vec{\alpha}}$$

$$\Delta dL_{k\vec{\alpha}} = -dM_{\vec{\alpha}} c^2 \Delta \left(\frac{1}{\gamma_{\vec{\alpha}}} \right) = dM_{\vec{\alpha}} \gamma_{\vec{\alpha}} \vec{v}_{\vec{\alpha}}(t) \cdot \frac{d\vec{\xi}_{\vec{\alpha}}(t)}{dt},$$

$$\Delta dL_{i\vec{\alpha}} = dQ_{\vec{\alpha}} \Big(\Delta \vec{A}(\vec{x}_{\vec{\alpha}}(t), t) \cdot \vec{v}_{\vec{\alpha}}(t) + \vec{A}(\vec{x}_{\vec{\alpha}}(t), t) \cdot \Delta \vec{v}_{\vec{\alpha}}(t)$$
$$- \Delta \phi(\vec{x}_{\vec{\alpha}}(t), t) \Big). \tag{37}$$

We can now combine the internal and kinetic parts of the varied Lagrangian taking into account the specific enthalpy definition given in equation (16):

$$\Delta dL_{k\vec{\alpha}} - \Delta dE_{in\ \vec{\alpha}} = dM_{\vec{\alpha}} \gamma_{\vec{\alpha}} \Big(\Big(1 + \frac{w_0}{c^2} \Big) \vec{v}_{\vec{\alpha}}(t) \cdot \frac{d\vec{\xi}_{\vec{\alpha}}(t)}{dt} + P_{\vec{\alpha}0} \vec{\nabla} \cdot \vec{\xi}_{\vec{\alpha}} \, dV_{\vec{\alpha},t}. \tag{38}$$

The electromagnetic interaction variation terms are not different than in the low speed (non-relativistic) case, see for example equations A47 and A48 of [1], and their derivation will not be repeated here:

$$d\vec{F}_{L\vec{\alpha}} \equiv dQ_{\vec{\alpha}} \left[\vec{v}_{\vec{\alpha}} \times \vec{B}(\vec{x}_{\vec{\alpha}}(t), t) + \vec{E}(\vec{x}_{\vec{\alpha}}(t), t) \right] \tag{39}$$

108

and:

$$\Delta dL_{i\vec{\alpha}} = \frac{d(dQ_{\vec{\alpha}}\vec{A}(\vec{x}_{\vec{\alpha}}(t),t)\cdot\vec{\xi}_{\vec{\alpha}})}{dt} + d\vec{F}_{L\vec{\alpha}}\cdot\vec{\xi}_{\vec{\alpha}}. \tag{40}$$

Let us introduce the shorthand notation:

$$\bar{\lambda} \equiv 1 + \frac{w_0}{c^2}, \qquad \lambda \equiv \gamma\bar{\lambda} = \gamma\left(1 + \frac{w_0}{c^2}\right). \tag{41}$$

If the enthalpy of a fluid element in its rest frame is much smaller than its rest energy:

$$dW_{\vec{\alpha}0} \ll dM_{\vec{\alpha}}c^2 \;\Rightarrow\; 1 \gg \frac{dW_{\vec{\alpha}0}}{dM_{\vec{\alpha}}c^2} = \frac{w_{\vec{\alpha}0}}{c^2}. \tag{42}$$

Thus in the classical limit (which involves restrictions on **both** the enthalpy of the fluid element and its velocity) we have:

$$\bar{\lambda} \simeq 1, \qquad \lambda \simeq 1. \tag{43}$$

The variation of the action of a relativistic fluid element is:

$$\Delta d\mathcal{A}_{\vec{\alpha}} = \int_{t1}^{t2}\Delta dL_{\vec{\alpha}}dt = (dM_{\vec{\alpha}}\lambda_{\vec{\alpha}}\vec{v}_{\vec{\alpha}}(t) + dQ_{\vec{\alpha}}\vec{A}(\vec{x}_{\vec{\alpha}}(t),t))\cdot\vec{\xi}_{\vec{\alpha}}\Big|_{t1}^{t2}$$
$$- \int_{t1}^{t2}(dM_{\vec{\alpha}}\frac{d(\lambda_{\vec{\alpha}}\vec{v}_{\vec{\alpha}}(t))}{dt}\cdot\vec{\xi}_{\vec{\alpha}} - d\vec{F}_{L\vec{\alpha}}\cdot\vec{\xi}_{\vec{\alpha}} - P_{\vec{\alpha}0}\vec{\nabla}\cdot\vec{\xi}_{\vec{\alpha}}\,dV_{\vec{\alpha},t})dt. \tag{44}$$

The variation of the relativistic fluid action is thus:

$$\Delta\mathcal{A} = \Delta\int_{\vec{\alpha}}d\mathcal{A}_{\vec{\alpha}} = \int_{\vec{\alpha}}(dM_{\vec{\alpha}}\lambda_{\vec{\alpha}}\vec{v}_{\vec{\alpha}}(t) + dQ_{\vec{\alpha}}\vec{A}(\vec{x}(\vec{\alpha},t),t))\cdot\vec{\xi}_{\vec{\alpha}}\Big|_{t1}^{t2}$$
$$- \int_{t1}^{t2}\int_{\vec{\alpha}}(dM_{\vec{\alpha}}\frac{d(\lambda_{\vec{\alpha}}\vec{v}_{\vec{\alpha}}(t))}{dt}\cdot\vec{\xi}_{\vec{\alpha}} - d\vec{F}_{L\vec{\alpha}}\cdot\vec{\xi}_{\vec{\alpha}} - P_{\vec{\alpha}0}\vec{\nabla}\cdot\vec{\xi}_{\vec{\alpha}}\,dV_{\vec{\alpha}})dt. \tag{45}$$

Now according to equation (4) we may write:

$$dM_{\vec{\alpha}} = \rho_{\vec{\alpha}}\,dV_{\vec{\alpha}}, \qquad dQ_{\vec{\alpha}} = \rho_{c\vec{\alpha}}\,dV_{\vec{\alpha}} \tag{46}$$

using the above relations we may turn the $\vec{\alpha}$ integral into a volume integral and thus write the variation of the fluid action in which we suppress the $\vec{\alpha}$ labels:

$$\Delta\mathcal{A} = \int(\rho\lambda\vec{v} + \rho_c\vec{A})\cdot\vec{\xi}dV\Big|_{t1}^{t2} - \int_{t1}^{t2}\int(\rho\frac{d(\lambda\vec{v})}{dt}\cdot\vec{\xi} - \vec{f}_L\cdot\vec{\xi} - P_0\vec{\nabla}\cdot\vec{\xi})dV\,dt. \tag{47}$$

in the above we introduced the Lorentz force density:

$$\vec{f}_{L\vec{\alpha}} \equiv \frac{d\vec{F}_{L\vec{\alpha}}}{dV_{\vec{\alpha}}} = \rho_{c\vec{\alpha}}\left[\vec{v}_{\vec{\alpha}}\times\vec{B}(\vec{x}_{\vec{\alpha}}(t),t) + \vec{E}(\vec{x}_{\vec{\alpha}}(t),t)\right]. \tag{48}$$

Now, since:

$$P_0\vec{\nabla}\cdot\vec{\xi} = \vec{\nabla}\cdot(P_0\vec{\xi}) - \vec{\xi}\cdot\vec{\nabla}P_0, \tag{49}$$

and using Gauss theorem the variation of the action can be written as:

$$\Delta \mathcal{A} = \int (\rho \lambda \vec{v} + \rho_c \vec{A}) \cdot \vec{\xi} dV \Big|_{t1}^{t2}$$
$$- \int_{t1}^{t2} \left[\int (\rho \frac{d(\lambda \vec{v})}{dt} - \vec{f}_L + \vec{\nabla} P_0) \cdot \vec{\xi} dV - \oint P_0 \vec{\xi} \cdot d\vec{\Sigma} \right] dt. \tag{50}$$

It follows that the variation of the action will vanish for a $\vec{\xi}$ such that $\vec{\xi}(t1) = \vec{\xi}(t2) = 0$ and vanishing on a surface encapsulating the fluid, but other than that arbitrary only if the Euler equation for a relativistic charged fluid is satisfied, that is:

$$\frac{d(\lambda \vec{v})}{dt} = -\frac{\vec{\nabla} P_0}{\rho} + \frac{\vec{f}_L}{\rho}, \qquad \frac{d\vec{v}}{dt} \simeq -\frac{\vec{\nabla} P_0}{\rho} + \frac{\vec{f}_L}{\rho}, \tag{51}$$

for the particular case that the fluid element is made of identical microscopic particles each with a mass m and a charge e, it follows that the mass and charge densities are proportional to the number density n:

$$\rho = m \, n, \quad \rho_c = e \, n \Rightarrow \frac{\vec{f}_L}{\rho} = k \left[\vec{v} \times \vec{B} + \vec{E} \right], k \equiv \frac{e}{m} \tag{52}$$

thus except from the terms related to the internal energy the equation is similar to that of a point particle. For a neutral fluid one obtains the form:

$$\frac{d(\lambda \vec{v})}{dt} = -\frac{\vec{\nabla} P_0}{\rho}, \qquad \frac{d\vec{v}}{dt} \simeq -\frac{\vec{\nabla} P_0}{\rho}. \tag{53}$$

Some authors prefer to write the above equation in terms of the energy per element of the fluid per unit volume in the rest frame which is the sum of the internal energy contribution and the rest mass contribution:

$$e_0 \equiv \rho_0 c^2 + \rho_0 \varepsilon_0. \tag{54}$$

It is easy to show that:

$$\bar{\lambda} = 1 + \frac{w_0}{c^2} = \frac{e_0 + P_0}{\rho_0 c^2}. \tag{55}$$

And using the above equality and some manipulations we may write equation (53) in a form which is preferable by some authors:

$$(e_0 + P_0) \frac{\gamma}{c^2} \frac{d(\gamma \vec{v})}{dt} = -\vec{\nabla} P_0 - \frac{\gamma^2}{c^2} \frac{dP_0}{dt} \vec{v}. \tag{56}$$

In practical fluid dynamics a fluid is described in terms of localized quantities, instead of quantities related to unseen infinitesimal "fluid elements". This is the Eulerian description of fluid dynamics in which one uses flow fields rather than "fluid elements" as will be discussed below. However, before we consider the Eulerian formalism a few words are in order of the canonical energy momentum tensor in general and in the particular description of the relativistic Lagrangian flow.

110

3 Canonical Energy-Momentum Tensor

Let us consider a general field of space-time of K components $\eta_a(x^\mu)$ and let us consider a Lagrangian density of the form:

$$\mathcal{L} = \mathcal{L}(\eta_a, \partial_\mu \eta_a), \qquad \partial_\mu \equiv \frac{\partial}{\partial x^\mu}. \tag{57}$$

Raising and lowering of indices will be done as usual using the Lorentz (Minkowski) metric $\eta_{\mu\nu} \equiv \text{diag}(+1, -1, -1, -1)$. The action:

$$\mathcal{A} \equiv \int d^4 x \mathcal{L} \tag{58}$$

will have an extremum such that $\delta \mathcal{A} = 0$ if $\mathcal{L}$ satisfies the Euler-Lagrange equations:

$$\frac{d}{dx_\nu} \left(\frac{\partial \mathcal{L}}{\partial (\partial_\nu \eta_a)} \right) = \frac{\partial \mathcal{L}}{\partial \eta_a}, \tag{59}$$

in the above $\frac{d}{dx_\nu}$ is the total derivative with respect to the coordinate x_ν, that is one takes into account the dependence of η_a on the coordinates. Also Einstein summation convention is assumed. The above is correct provided that the variation is fixed and the initial and final times and the boundary conditions are satisfied.

Then one can define an energy momentum tensor of the form [10]:

$$T_{\mu\nu} \equiv \partial_\mu \eta_a \frac{\partial \mathcal{L}}{\partial (\partial^\nu \eta_a)} - \eta_{\mu\nu} \mathcal{L}. \tag{60}$$

In the above the Einstein summation convention is assumed with respect to the scalar field index a. This quantity will satisfy due to Euler-Lagrange equations (59) the following conservation laws:

$$\frac{d}{dx_\nu} T_{\mu\nu} = 0, \tag{61}$$

provided $\mathcal{L}$ does not depend explicitly on the coordinates as is appropriate for any $\mathcal{L}$ describing universal physical laws.

The quantities:

$$\bar{P}_\mu = \int d^3 x T_{\mu 0} \tag{62}$$

are conserved in time, that is:

$$\frac{d\bar{P}_\mu}{dt} = 0. \tag{63}$$

$\bar{P}_0$ is the total energy of the system and $\bar{P}_i$ are the components of the total linear momentum of the system $i \in [1, 2, 3]$.

For a Lagrangian fluid there is dependence of one coordinate t ($x^0 = ct$), however, the field indices are continuous (through the $\vec{\alpha}$ label) and therefore infinite. It follows that:

$$T_{L00} = \int_{\vec{\alpha}} \left[\frac{d\vec{x}_{\vec{\alpha}}(t)}{dt} \cdot \frac{\partial L}{\partial(\frac{d\vec{x}_{\vec{\alpha}}(t)}{dt})} \right] - L = \int_{\vec{\alpha}} \left[\vec{v}_{\vec{\alpha}} \cdot \frac{\partial L}{\partial \vec{v}_{\vec{\alpha}}} \right] - L. \tag{64}$$

111

Now according to equation (3):

$$\frac{\partial L}{\partial \vec{v}_{\vec{\alpha}}} = \frac{\partial \int_{\vec{\alpha}'} dL_{\vec{\alpha}'}}{\partial \vec{v}_{\vec{\alpha}}} = \frac{\partial dL_{\vec{\alpha}}}{\partial \vec{v}_{\vec{\alpha}}}. \tag{65}$$

Thus it follows from equation (2) that:

$$\frac{\partial L}{\partial \vec{v}_{\vec{\alpha}}} = \frac{\partial dL_{k\vec{\alpha}}}{\partial \vec{v}_{\vec{\alpha}}} + \frac{\partial dL_{i\vec{\alpha}}}{\partial \vec{v}_{\vec{\alpha}}} - \frac{\partial dE_{in\ \vec{\alpha}}}{\partial \vec{v}_{\vec{\alpha}}}. \tag{66}$$

Using the definitions of $dL_{k\vec{\alpha}}, dL_{i\vec{\alpha}}, dE_{in\ \vec{\alpha}}$ given in equation (2), and the fact that $dE_{in\ \vec{\alpha}}$ is velocity dependent according to equation (24) it follows that:

$$\frac{\partial L}{\partial \vec{v}_{\vec{\alpha}}} = dM_{\vec{\alpha}} \gamma_{\vec{\alpha}} \vec{v}_{\vec{\alpha}} + dQ_{\vec{\alpha}} \vec{A}_{\vec{\alpha}} + dE_{in\ \vec{\alpha}0} \frac{\gamma_{\vec{\alpha}}}{c^2} \vec{v}_{\vec{\alpha}}. \tag{67}$$

in which we have used the identity:

$$\frac{\partial \gamma_{\vec{\alpha}}^{-1}}{\partial \vec{v}_{\vec{\alpha}}} = -\frac{\gamma_{\vec{\alpha}}}{c^2} \vec{v}_{\vec{\alpha}}. \tag{68}$$

Plugging equation (68) into equation (64) will lead after some simplification to the form:

$$T_{L00} = \int_{\vec{\alpha}} \left[dM_{\vec{\alpha}} c^2 \gamma_{\vec{\alpha}} + dQ_{\vec{\alpha}} \phi(\vec{x}_{\vec{\alpha}}) + dE_{in\ \vec{\alpha}0} \gamma_{\vec{\alpha}} \right]. \tag{69}$$

This can be written as a volume integral:

$$T_{L00} = \int d^3 x \left[\rho(c^2 + \varepsilon_0)\gamma + \rho_c \phi \right]. \tag{70}$$

The above expression is the total energy of the fluid which is conserved.

4 The Action Principle of a Relativistic Charged Eulerian Fluid

Here we follow the derivation of [1, 2, 13] but now taking into account the relativistic nature of the flow, this implies taking into account an action which is invariant under Lorentz transformations. Let us consider the action:

$$\mathcal{A} \equiv \int \mathcal{L} d^3 x dt, \qquad \mathcal{L} \equiv \mathcal{L}_0 + \mathcal{L}_2 + \mathcal{L}_i$$

$$\mathcal{L}_0 \equiv -\rho(\frac{c^2}{\gamma} + \varepsilon) = -\rho_0(c^2 + \varepsilon_0) = -e_0,$$

$$\mathcal{L}_2 \equiv \nu \partial^\nu (\rho_0 u_\nu) - \rho_0 \alpha u_\nu \partial^\nu \beta - \rho_0 \sigma u_\nu \partial^\nu s,$$

$$\mathcal{L}_i \equiv -\rho_c A^\nu v_\nu, \qquad v_\nu \equiv \frac{dx_\nu}{dt}. \tag{71}$$

In the non relativistic limit we may write:

$$\mathcal{L}_0 \simeq \rho(\frac{1}{2}v^2 - \varepsilon - c^2) \tag{72}$$

Taking into account that:

$$u_\mu = \gamma(c, \vec{v}) \tag{73}$$

and also that $\rho = \gamma\rho_0$ according to equation (23), it is easy to write the above Lagrangian densities in a space-time formalism:

$$\mathcal{L}_2 = \nu[\frac{\partial\rho}{\partial t} + \vec{\nabla}\cdot(\rho\vec{v})] - \rho\alpha\frac{d\beta}{dt} - \rho\sigma\frac{ds}{dt} \qquad \mathcal{L}_i = \rho_c\left(\vec{A}\cdot\vec{v} - \phi\right). \tag{74}$$

Here we consider the variational variables to be functions of space and time (fields). Those include a vector velocity field $\vec{v}(\vec{x}, t)$ and density scalar field $\rho(\vec{x}, t)$. In the above we use the material derivative defined by the prevalent form:

$$\frac{dg(\vec{\alpha}, t)}{dt} = \frac{dg(\vec{x}(\vec{\alpha}, t), t)}{dt} = \frac{\partial g}{\partial t} + \frac{d\vec{x}}{dt}\cdot\vec{\nabla}g = \frac{\partial g}{\partial t} + \vec{v}\cdot\vec{\nabla}g \tag{75}$$

once g is taken to be dependent on $\vec{x}, t$. The way to include in the above formalism conservation of quantities which is easy in the Lagrange approach such as: label, mass, charge and entropy is by using Lagrange multipliers ν, α, σ that enforce the equations:

$$\frac{\partial\rho}{\partial t} + \vec{\nabla}\cdot(\rho\vec{v}) = 0$$
$$\frac{d\beta}{dt} = 0$$
$$\frac{ds}{dt} = 0. \tag{76}$$

Provided $\rho \neq 0$ those are the continuity equation which ensures mass conservation and the conditions that β is a label (comoving function), and the specific entropy s is conserved. Combining variation with respect to β with the continuity equation (76) will lead to the equation (similar to the derivation of equations (67-69) of [2]):

$$\frac{d\alpha}{dt} = 0 \tag{77}$$

Hence for $\rho \neq 0$ both α and β are labels. The specific internal energy ε_0 defined in equation (6) is dependent on the thermodynamic properties of the fluid. This is formulated through an equation of state as a function of the density and specific entropy. We assume that the entropy of a fluid element is Lorentz invariant that is $dS_{\vec{\alpha}0} = dS_{\vec{\alpha}}$ and hence $s_{\vec{\alpha}0} = s_{\vec{\alpha}}$. Varying the action with respect to the specific entropy s we obtain:

$$\frac{d\sigma}{dt} = \frac{T_0}{\gamma}, \tag{78}$$

provided that the appropriate boundary conditions are satisfied. The electromagnetic potentials $\vec{A}, \phi$ are given functions of coordinates and thus are not varied. Another assumption in our analysis is that each fluid element is composed of microscopic particles of mass m and charge e, thus it follows from equation (52) that:

$$\rho_c = k\rho. \tag{79}$$

Let us now take the variational derivative with respect to the density ρ, we obtain:

$$
\begin{aligned}
\delta_\rho A &= \int d^3x dt \delta \rho [-\frac{c^2}{\gamma} - w_0 \frac{\delta \rho_0}{\delta \rho} - \frac{\partial \nu}{\partial t} - \vec{v} \cdot \vec{\nabla} \nu + k(\vec{A} \cdot \vec{v} - \phi)] \\
&+ \oint d\vec{S} \cdot \vec{v} \delta \rho \nu + \int d\vec{\Sigma} \cdot \vec{v} \delta \rho [\nu] + \int d^3x \nu \delta \rho |_{t_0}^{t_1}
\end{aligned} \tag{80}
$$

Or as:

$$
\begin{aligned}
\delta_\rho A &= \int d^3x dt \delta \rho [-\frac{c^2 + w_0}{\gamma} - \frac{\partial \nu}{\partial t} - \vec{v} \cdot \vec{\nabla} \nu + k(\vec{A} \cdot \vec{v} - \phi)] \\
&+ \oint d\vec{S} \cdot \vec{v} \delta \rho \nu + \int d\vec{\Sigma} \cdot \vec{v} \delta \rho [\nu] + \int d^3x \nu \delta \rho |_{t_0}^{t_1}
\end{aligned} \tag{81}
$$

Hence if $\delta \rho$ disappears on the boundary and cut, and in initial and final times we obtain:

$$\frac{d\nu}{dt} = \frac{\partial \nu}{\partial t} + \vec{v} \cdot \vec{\nabla} \nu = -\frac{c^2 + w_0}{\gamma} + k(\vec{A} \cdot \vec{v} - \phi) \Rightarrow \frac{d\nu}{dt} \simeq \frac{1}{2} v^2 - w_0 - c^2 + k(\vec{A} \cdot \vec{v} - \phi). \tag{82}$$

Equation (82) can be written in the short form:

$$\frac{d\nu}{dt} = -c^2 \frac{\bar{\lambda}}{\gamma} + k(\vec{A} \cdot \vec{v} - \phi) = -c^2 \frac{\lambda}{\gamma^2} + k(\vec{A} \cdot \vec{v} - \phi), \tag{83}$$

in which $\bar{\lambda}$ is defined in equation (41). Finally we vary the action with respect to $\vec{v}$, taking into account that:

$$\delta_{\vec{v}} \frac{1}{\gamma} = -\gamma \frac{\vec{v} \cdot \delta \vec{v}}{c^2} \tag{84}$$

This will result in:

$$
\begin{aligned}
\delta_{\vec{v}} A &= \int d^3x dt \rho \delta \vec{v} \cdot [\gamma \vec{v} - \frac{w_0}{\rho} \frac{\delta \rho_0}{\delta \vec{v}} - \vec{\nabla} \nu - \alpha \vec{\nabla} \beta - \sigma \vec{\nabla} s + k\vec{A}] \\
&+ \oint d\vec{S} \cdot \delta \vec{v} \rho \nu + \int d\vec{\Sigma} \cdot \delta \vec{v} \rho [\nu].
\end{aligned} \tag{85}
$$

However:

$$\frac{\delta \rho_0}{\delta \vec{v}} = \rho \frac{\delta \frac{1}{\gamma}}{\delta \vec{v}} = -\rho \gamma \frac{\vec{v}}{c^2} \tag{86}$$

Taking in account the definition of λ (see equation (41)), we thus have:

$$
\begin{aligned}
\delta_{\vec{v}} A &= \int d^3x dt \rho \delta \vec{v} \cdot [\lambda \vec{v} - \vec{\nabla} \nu - \alpha \vec{\nabla} \beta - \sigma \vec{\nabla} s + k\vec{A}] \\
&+ \oint d\vec{S} \cdot \delta \vec{v} \rho \nu + \int d\vec{\Sigma} \cdot \delta \vec{v} \rho [\nu].
\end{aligned} \tag{87}
$$

114

the above boundary terms contain integration over the external boundary $\oint d\vec{S}$ and an integral over the cut $\int d\vec{\Sigma}$ that must be introduced in case that ν is not single valued. The external boundary term vanishes; in the case of astrophysical flows for which $\rho = 0$ on the free flow boundary, or the case in which the fluid is contained in a vessel which induces a no flux boundary condition $\delta\vec{v} \cdot \hat{n} = 0$ ($\hat{n}$ is a unit vector normal to the boundary). The cut "boundary" term vanish when the velocity field varies only parallel to the cut that is it satisfies a Kutta type condition. If the boundary terms vanish $\vec{v}$ must have the following form:

$$\lambda\vec{v} = \alpha\vec{\nabla}\beta + \sigma\vec{\nabla}s + \vec{\nabla}\nu - k\vec{A}, \Rightarrow \vec{v} \simeq \alpha\vec{\nabla}\beta + \sigma\vec{\nabla}s + \vec{\nabla}\nu - k\vec{A}, \tag{88}$$

this is a generalization of Clebsch representation of a non-barotropic flow field (see for example [11], [14, page 248]) for a relativistic charged flow. We notice that the ν function contributes to the velocity field through its gradient, this means that we can add any function of time to ν without altering the physical field. Redefining:

$$\bar{\nu} \equiv \nu + c^2 t \tag{89}$$

we obtain the more standard classical form of equation (82) and equation (88).

$$\frac{d\bar{\nu}}{dt} \simeq \frac{1}{2}v^2 - w_0 + k(\vec{A} \cdot \vec{v} - \phi), \qquad \vec{v} \simeq \alpha\vec{\nabla}\beta + \sigma\vec{\nabla}s + \vec{\nabla}\bar{\nu} - k\vec{A}. \tag{90}$$

4.1   Euler's equations

We shall now show that a velocity field given by equation (88), such that the functions α, β, ν satisfy the corresponding equations (76,82,77) must satisfy Euler's equations. Let us calculate the material derivative of $\lambda\vec{v}$:

$$\frac{d(\lambda\vec{v})}{dt} = \frac{d\vec{\nabla}\nu}{dt} + \frac{d\alpha}{dt}\vec{\nabla}\beta + \alpha\frac{d\vec{\nabla}\beta}{dt} + \frac{d\sigma}{dt}\vec{\nabla}s + \sigma\frac{d\vec{\nabla}s}{dt} - k\frac{d\vec{A}}{dt} \tag{91}$$

It can be easily shown that:

$$\begin{aligned}
\frac{d\vec{\nabla}\nu}{dt} &= \vec{\nabla}\frac{d\nu}{dt} - \vec{\nabla}v_n\frac{\partial\nu}{\partial x_n} = \vec{\nabla}\left(-\frac{c^2 + w_0}{\gamma} + k\vec{A} \cdot \vec{v} - k\phi\right) - \vec{\nabla}v_n\frac{\partial\nu}{\partial x_n} \\
\frac{d\vec{\nabla}\beta}{dt} &= \vec{\nabla}\frac{d\beta}{dt} - \vec{\nabla}v_n\frac{\partial\beta}{\partial x_n} = -\vec{\nabla}v_n\frac{\partial\beta}{\partial x_n} \\
\frac{d\vec{\nabla}s}{dt} &= \vec{\nabla}\frac{ds}{dt} - \vec{\nabla}v_n\frac{\partial s}{\partial x_n} = -\vec{\nabla}v_n\frac{\partial s}{\partial x_n}
\end{aligned} \tag{92}$$

115

In which x_n is a Cartesian coordinate and a summation convention is assumed. Inserting the result from equations (92) into equation (91) yields:

$$
\begin{aligned}
\frac{d(\lambda \vec{v})}{dt} &= -\vec{\nabla} v_n \left(\frac{\partial \nu}{\partial x_n} + \alpha \frac{\partial \beta}{\partial x_n} + \sigma \frac{\partial s}{\partial x_n} \right) + \vec{\nabla} \left(-\frac{c^2 + w_0}{\gamma} + k\vec{A} \cdot \vec{v} - k\phi \right) \\
&\quad - k\frac{d\vec{A}}{dt} + \frac{T_0}{\gamma} \vec{\nabla} s \\
&= -\vec{\nabla} v_n (\lambda v_n + k A_n) + \vec{\nabla} \left(-\frac{c^2 + w_0}{\gamma} + k\vec{A} \cdot \vec{v} - k\phi \right) \\
&\quad - k\partial_t \vec{A} - k(\vec{v} \cdot \vec{\nabla})\vec{A} + \frac{T_0}{\gamma} \vec{\nabla} s \\
&= -\frac{1}{\gamma} \vec{\nabla} w_0 + k\vec{E} + k(v_n \vec{\nabla} A_n - v_n \partial_n \vec{A}) + \frac{T_0}{\gamma} \vec{\nabla} s,
\end{aligned}
\tag{93}
$$

the electric field is defined in equation (8) of [5]. Also according to equation (7) of [5]:

$$
(v_n \vec{\nabla} A_n - v_n \partial_n \vec{A})_l = v_n (\partial_l A_n - \partial_n A_l) = \epsilon_{lnj} v_n B_j = (\vec{v} \times \vec{B})_l, \tag{94}
$$

Hence we obtain the Euler equation of a charged relativistic fluid in the form:

$$
\frac{d(\lambda \vec{v})}{dt} = -\frac{1}{\gamma} \vec{\nabla} w_0 + \frac{T_0}{\gamma} \vec{\nabla} s + k \left[\vec{v} \times \vec{B} + \vec{E} \right] = -\frac{1}{\rho} \vec{\nabla} P_0 + k \left[\vec{v} \times \vec{B} + \vec{E} \right], \tag{95}
$$

in which we have used equation (19). Equation (95) is identical to equation (51) and thus it is proven that the Euler equations can be derived from the action (74) thus all the equations of relativistic charged fluid dynamics can be derived from the action (74) using arbitrary and unrestricted variations.

4.2   Simplified action

It may be claimed the previous approach introduced unnecessary complications to the theory of relativistic fluid dynamics by adding additional four more scalar fields $\alpha, \beta, \nu, \sigma$ to the physical set $\vec{v}, \rho, s$. We will show that this is just a superficial impression and equation (71) given in a pedagogical form can be simplified. It is easy to show that defining a four dimensional non-barotropic Clebsch four vector:

$$
\begin{aligned}
v_C^\mu &\equiv \alpha \partial^\mu \beta + \sigma \partial^\mu s + \partial^\mu \nu = (\frac{1}{c}(\alpha \partial_t \beta + \sigma \partial_t s + \partial_t \nu), \alpha \vec{\nabla} \beta + \sigma \vec{\nabla} s + \vec{\nabla} \nu) \\
&= (\frac{1}{c}(\alpha \partial_t \beta + \sigma \partial_t s + \partial_t \nu), \vec{v}_C)
\end{aligned}
\tag{96}
$$

and a four dimensional non-barotropic electromagnetic Clebsch four vector:

$$
v_E^\mu \equiv v_C^\mu + k A^\mu = (v_E^0, \vec{v}_E) = (\frac{1}{c}(\alpha \partial_t \beta + \sigma \partial_t s + \partial_t \nu + k\phi), \vec{v}_C - k\vec{A}). \tag{97}
$$

It follows from equation (76), equation (83) and equation (88) that:

$$
v_\mu = -\frac{v_{E\mu}}{\lambda} \quad \Rightarrow \quad \vec{v} = \frac{\vec{v}_E}{\lambda}, \quad \lambda = -\frac{1}{c^2}(\alpha \partial_t \beta + \sigma \partial_t s + \partial_t \nu + k\phi). \tag{98}
$$

116

The classical limit of the above equations is:

$$\vec{v} \simeq \vec{v}_E, \quad \lambda \simeq 1. \tag{99}$$

Eliminating $\vec{v}$ the Lagrangian density appearing in equation (74) can be written (up to surface terms) in the compact form:

$$\hat{\mathcal{L}}[\rho_0, s, \alpha, \beta, \nu, \sigma] = \rho_0 \left[c \sqrt{v_{E\mu} v_E^\mu} - \varepsilon_0 - c^2 \right] \tag{100}$$

This Lagrangian density will yield the six equations (76,77,78,82), after those equations are solved we can substitute the scalar fields $\alpha, \beta, s, \sigma, \nu$ into equation (88) to obtain $\vec{v}$. Hence, the general charged relativistic non-barotropic fluid dynamics problem is modified such that instead of solving the Euler and continuity equations we need to solve an alternative equivalent set which can be derived from the Lagrangian density $\hat{\mathcal{L}}$. The classical limit of the equation (100) can be calculated as follows. First notice that:

$$\sqrt{v_{E\mu} v_E^\mu} = |v_{E0}| \sqrt{1 - \frac{\vec{v}_E^2}{v_{E0}^2}} = |v_{E0}| \sqrt{1 - \frac{\vec{v}^2}{c^2}} = \frac{|v_{E0}|}{\gamma} = \frac{c|\lambda|}{\gamma} = c|\bar{\lambda}| \tag{101}$$

in which we used equation (98). According to equation (98) $v_{E0} < 0$ as $\lambda > 0$, and thus :

$$|v_{E0}| = -v_{E0} = -\frac{1}{c}(\alpha \partial_t \beta + \sigma \partial_t s + \partial_t \nu + k\phi). \tag{102}$$

Combining equation (101) and equation (102) it follows that:

$$c \sqrt{v_{E\mu} v_E^\mu} = -\frac{1}{\gamma}(\alpha \partial_t \beta + \sigma \partial_t s + \partial_t \nu + k\phi) \tag{103}$$

Which can be written in term of $\bar{\nu}$, defined in equation (89) as:

$$c \sqrt{v_{E\mu} v_E^\mu} = -\frac{1}{\gamma}(\alpha \partial_t \beta + \sigma \partial_t s + \partial_t \bar{\nu} + k\phi - c^2) \tag{104}$$

We can now write equation (100) in the form:

$$\begin{aligned} \hat{\mathcal{L}} &= \rho_0 \left[-\frac{1}{\gamma}(\alpha \partial_t \beta + \sigma \partial_t s + \partial_t \bar{\nu} + k\phi - c^2) - \varepsilon_0 - c^2 \right] \\ &= -\frac{\rho_0}{\gamma} \left[\alpha \partial_t \beta + \sigma \partial_t s + \partial_t \bar{\nu} + k\phi + \gamma \varepsilon_0 + (\gamma - 1)c^2 \right]. \end{aligned} \tag{105}$$

The classical limit is straight forward: $\frac{v}{c} \to 0, \gamma \to 1$, except for the last term which should be handled with care:

$$(\gamma - 1)c^2 = \left(\frac{\gamma - 1}{\left(\frac{v}{c}\right)^2} \right) v^2 \simeq \frac{1}{2}v^2 \simeq \frac{1}{2}v_E^2. \tag{106}$$

Taking into account equation (106) it follows that:

$$\hat{\mathcal{L}} \simeq -\rho_0 [\frac{\partial \bar{\nu}}{\partial t} + \alpha \frac{\partial \beta}{\partial t} + \sigma \frac{\partial s}{\partial t} + \varepsilon(\rho_0, s_0) + k\phi + \frac{1}{2}(\vec{\nabla}\bar{\nu} + \alpha\vec{\nabla}\beta + \sigma\vec{\nabla}s - k\vec{A})^2] \qquad (107)$$

However, from equation (24) it follows that:

$$\varepsilon_0 = \frac{dE_{in\ \bar{\alpha}0}}{dM_{\bar{\alpha}}} = \gamma \frac{dE_{in\ \bar{\alpha}}}{dM_{\bar{\alpha}}} = \gamma\varepsilon. \qquad (108)$$

Thus in the classical limit there is no difference between ρ and ρ_0, and ε_0 and ε, since the difference is of order $\left(\frac{v}{c}\right)^2$. Therefore we may write:

$$\hat{\mathcal{L}} \simeq -\rho [\frac{\partial \bar{\nu}}{\partial t} + \alpha \frac{\partial \beta}{\partial t} + \sigma \frac{\partial s}{\partial t} + \varepsilon(\rho, s) + k\phi + \frac{1}{2}(\vec{\nabla}\bar{\nu} + \alpha\vec{\nabla}\beta + \sigma\vec{\nabla}s - k\vec{A})^2] \qquad (109)$$

5   Energy Momentum Tensor

We shall now calculate the canonical energy momentum tensor of the Lagrangian density given in equation (100), this is done through the definition given in equation (60). However, as $\hat{\mathcal{L}}$ is dependent only on the derivatives: $\partial^\mu \beta, \partial^\mu s, \partial^\mu \nu$ we obtain:

$$T_{E\mu\nu} = \partial_\mu \beta \frac{\partial \hat{\mathcal{L}}}{\partial(\partial^\nu \beta)} + \partial_\mu s \frac{\partial \hat{\mathcal{L}}}{\partial(\partial^\nu s)} + \partial_\mu \nu \frac{\partial \hat{\mathcal{L}}}{\partial(\partial^\nu \nu)} - \eta_{\mu\nu}\hat{\mathcal{L}}. \qquad (110)$$

Taking into account equation (101) it is easy to see that:

$$\frac{\partial \hat{\mathcal{L}}}{\partial(\partial^\nu \beta)} = \frac{\rho\alpha}{\lambda} v_{E\nu}, \qquad \frac{\partial \hat{\mathcal{L}}}{\partial(\partial^\nu s)} = \frac{\rho\sigma}{\lambda} v_{E\nu}, \qquad \frac{\partial \hat{\mathcal{L}}}{\partial(\partial^\nu \nu)} = \frac{\rho}{\lambda} v_{E\nu}. \qquad (111)$$

Plugging equation (111) into equation (110) we have:

$$T_{E\mu\nu} = \frac{\rho}{\lambda} \left(\alpha\partial_\mu \beta + \sigma\partial_\mu s + \partial_\mu \nu\right) v_{E\nu} - \eta_{\mu\nu}\hat{\mathcal{L}} = \frac{\rho}{\lambda} v_{C\mu} v_{E\nu} - \eta_{\mu\nu}\hat{\mathcal{L}}, \qquad (112)$$

in which we have used the definition given in equation (96). The above expression can be written in terms of more physical velocity fields using equation (97) and equation (98) as:

$$T_{E\mu\nu} = \rho(\lambda v_\mu + kA_\mu)v_\nu - \eta_{\mu\nu}\hat{\mathcal{L}}. \qquad (113)$$

By using equation (101) we notice that the Lagrangian density of equation (100) can be written as:

$$\hat{\mathcal{L}} = \rho_0 \left[c^2\bar{\lambda} - \varepsilon_0 - c^2\right]. \qquad (114)$$

Taking into account the definition of $\bar{\lambda}$ given in equation (41) we see that $\hat{\mathcal{L}}$ is numerically equal to the pressure at the rest frame:

$$\hat{\mathcal{L}} = \rho_0 \left[c^2(1 + \frac{w_0}{c^2}) - \varepsilon_0 - c^2\right] = \rho_0 \left[w_0 - \varepsilon_0\right] = P_0. \qquad (115)$$

in which for the last equality sign we have used equation (17). Thus:

$$T_{E\mu\nu} = \rho(\lambda v_\mu + kA_\mu)v_\nu - \eta_{\mu\nu}P_0. \tag{116}$$

It is interesting to compare this expression to the one suggested by Weinberg [7], to do this we look at uncharged fluids for which $k = 0$. Using the definition of λ given in equation (41) and equation (73) we obtain:

$$\rho\lambda v_\mu v_\nu = \rho\lambda\frac{u_\mu u_\nu}{\gamma^2} = \rho_0\bar{\lambda}u_\mu u_\nu = \rho_0(1 + \frac{w_0}{c^2})u_\mu u_\nu = \rho_0(1 + \frac{\varepsilon_0 + \frac{P_0}{\rho_0}}{c^2})u_\mu u_\nu \tag{117}$$

Thus:

$$\rho\lambda v_\mu v_\nu = \frac{1}{c^2}(\rho_0 c^2 + \rho_0\varepsilon_0 + P_0)u_\mu u_\nu = \frac{1}{c^2}(e_0 + P_0)u_\mu u_\nu \tag{118}$$

in which we have used the definition of energy per element of the fluid per unit volume in the rest frame given in equation (54). Thus we may rewrite the energy momentum tensor of an uncharged flow in the form:

$$T_{E\mu\nu} = \frac{1}{c^2}(e_0 + P_0)u_\mu u_\nu - \eta_{\mu\nu}P_0. \tag{119}$$

This should be compared to the Weinberg [7] form:

$$T_{W\mu\nu} = (\rho_W + P_W)u_{W\mu}u_{W\nu} + \eta_{W\mu\nu}P_W. \tag{120}$$

In which we need to take into account that Weinberg does not use practical MKS units as we do, but natural units in which $c = 1$, thus $u_{W\mu} = u_\mu$. The choice of metric that Weinberg uses is also opposite to our own hence $\eta_{W\mu\nu} = -\eta_{\mu\nu}$. It also follows that Weinberg's pressure is the pressure at the rest frame $P_w = P_0$ and that Weinberg's proper energy density is the total energy density in the rest frame: $\rho_w = e_0$.

6 Conclusion

Current cosmological models in particular ΛCDM postulate an unconfirmed component (the Λ in ΛCDM) which is denoted "dark energy". This is needed to fit the accelerating expansion of the universe as obtained through distant supernovae observations, the CMB spectrum and the large scale structure which developed through baryonic acoustic oscillations in the primordial matter. The energy momentum tensor obtain in equation (116) allows us two alternative routes. One is to assume that primordial fluid was not electrically neutral and thus one must take into account the electromagnetic contributions to the energy-momentum tensor. The other one is to take into account a non barotropic fluid in which the quantity $\rho_0\varepsilon_0$ is not just a simple multiplication of P_0, which may be either 3 or $\frac{3}{2}$ depending on wether the fluid is ultra relativistic or non-relativistic. But rather a more realistic form in which $\rho_0\varepsilon_0$ is dependent on two thermodynamical quantities which can be either the density, specific entropy, pressure and temperature all defined in the fluid element's rest frame. Of course more detailed analysis is needed which will be hopefully published in future papers.

References

[1] Yahalom, A. Fisher Information Perspective of Pauli's Electron. Entropy 2022, 24, 1721. https://doi.org/10.3390/e24121721.

[2] Yahalom, A. (2023). Fisher Information Perspective of Pauli's Electron. In: Skiadas, C.H., Dimotikalis, Y. (eds) 15th Chaotic Modeling and Simulation International Conference. CHAOS 2022. Springer Proceedings in Complexity. Springer, Cham. https://doi.org/10.1007/978-3-031-27082-6_26

[3] Clebsch, A. 1857 Uber eine allgemeine transformation der hydro-dynamischen Gleichungen. J. Reine Angew. Math. 54, 293-312.

[4] Clebsch, A. 1859 Uber die Integration der hydrodynamischen Gleichungen. J. Reine Angew. Math. 56, 1-10.

[5] Yahalom, Asher. 2023. "A Fluid Perspective of Relativistic Quantum Mechanics" Entropy 25, no. 11: 1497. https://doi.org/10.3390/e25111497

[6] Yahalom, A. Dirac Equation and Fisher Information. Entropy 2024, 26, 971. https://doi.org/10.3390/e26110971

[7] Weinberg, S. *Gravitation and Cosmology: Principles and Applications of the General Theory of Relativity*; John Wiley & Sons, Inc.: Hoboken, NJ, USA, 1972.

[8] Misner, C.W.; Thorne, K.S.; Wheeler, J.A. *Gravitation*; W.H. Freeman & Company: New York, NY, USA, 1973.

[9] Padmanabhan, T. *Gravitation: Foundations and Frontiers*; Cambridge University Press: Cambridge, UK, 2010.

[10] H. Goldstein , C. P. Poole Jr. & J. L. Safko, Classical Mechanics, Pearson; 3 edition (2001).

[11] C. Eckart, "Variation Principles of Hydrodynamics," *Phys. Fluids*, vol. 3, 421, 1960.

[12] F.P. Bretherton "A note on Hamilton's principle for perfect fluids," Journal of Fluid Mechanics / Volume 44 / Issue 01 / October 1970, pp 19 31 DOI: 10.1017/S0022112070001660, Published online: 29 March 2006.

[13] A. Yahalom "The Fluid Dynamics of Spin". Molecular Physics, Published online: 13 Apr 2018. http://dx.doi.org/10.1080/00268976.2018.1457808 (arXiv:1802.09331 [physics.flu-dyn]).

[14] Lamb, H., *Hydrodynamics*, 1945 (New York: Dover Publications).

[15] A. Yahalom "The Fluid Dynamics of Spin - a Fisher Information Perspective" arXiv:1802.09331v2 [cond-mat.] 6 Jul 2018. Proceedings of the Seventeenth Israeli - Russian Bi-National Workshop 2018 "The optimization of composition, structure and properties of metals, oxides, composites, nano and amorphous materials".

[16] Asher Yahalom "The Fluid Dynamics of Spin - a Fisher Information Perspective and Comoving Scalars" Chaotic Modeling and Simulation (CMSIM) 1: 17-30, 2020.

[17] Gronwald, Frank and Muench, Uwe and Macías, Alfredo and Hehl, Friedrich W. "Volume elements of spacetime and a quartet of scalar fields", Phys. Rev. D, 58, 8, 084021, 1998, American Physical Society, https://link.aps.org/doi/10.1103/PhysRevD.58.084021, e-Print: gr-qc/9712063.

[18] E. I. Guendelman and A. B. Kaganovich, Phys. Rev. D53, 7020 (1996), gr-qc/9605026.

[19] E. I. Guendelman and A. B. Kaganovich, Phys. Rev. D55, 5970 (1997), gr-qc/9611046.

[20] E. I. Guendelman and A. B. Kaganovich, Phys. Rev. D56, 3548 (1997), gr-qc/9702058.

6 THE POST-PARADIGMATIC CHARACTER OF CONTEMPORARY COSMOLOGY

ANGUEL S. STEFANOV

Abstract   This paper pursues the achievement of two aims. The first one is the claim that in contrast to all other contemporary scientific disciplines, which have long ago entered into their paradigmatic stage of development according to the methodological conception of Thomas Kuhn, cosmology has passed along another path of its growth. It has went out of its classical paradigmatic stage and today it is in a post-paradigmatic one. My second aim is the thesis that the ontology of contemporary cosmological theories could be affected by philosophical and aesthetical predilections, which although being set as not exactly scientific kind of prerequisites, exert methodological influence on scientific theorization.

Keywords: classical cosmological paradigm, post-paradigmatic stage, universe, cosmological theories, recent astronomical observations.

This paper pursues the achievement of two aims. The first one is the claim that, in contrast to all other contemporary scientific disciplines, which have long ago entered into their paradigmatic stage of development according to the methodological conception of Thomas Kuhn, cosmology has passed along another path of its growth. It has went out of its paradigmatic stage and today it is in a post-paradigmatic one. My second aim is the thesis that the construction of contemporary cosmological theories could be affected by philosophical and aesthetical predilections, which although being set as not exactly scientific kind of prerequisites, exert methodological influence on scientific theorization.

I shall hardly astonish the reader by saying that cosmology is probably the most intensively growing discipline nowadays. This is, generally speaking, due to the human curiosity to grasp what is the nature of the universe and the place of man in it. Nevertheless, it is mostly because of the successful interplay among different spheres of contemporary scientific knowledge, involving models of the constitution of the micro-world, astronomical and astrophysical knowledge about the formation and the dynamic behavior of stars and galaxies, as well as an understanding of the nature of spacetime. All this goes hand in hand with the application of experimental devices of high precision (space telescopes, huge detectors of gravitational waves, and the like). This entire contemporary knowledge has pushed cosmology out of its paradigmatic stage into an ensemble of different theories about the genesis and the features of our universe.

Until very recently, having in mind the span of human history, say till the end of the twenties of the last century, our notion of the whole universe was limited only by the observations of our own galaxy – the Milky Way, and some accompanying nebulae. Andromeda for instance was estimated to be a nebula of gas, not a galaxy. Only due to the investigations of Edwin Hubble and Knut Lundmark astronomers came to know that there are some other galaxies beyond the Milky Way, like Andromeda and the Magellan clouds; and we are sure

Kyley Ewing (Ed.), *Spacetime Conference 2024. Selected peer-reviewed papers presented at the Seventh International Conference on the Nature and Ontology of Spacetime, 16 - 19 September 2024, Albena, Bulgaria* (Minkowski Institute Press, Montreal 2025). ISBN 978-1-998902-44-6 (softcover), ISBN 978-1-998902-45-3 (ebook).

today that there are billions of them.

The discovery of other galaxies did not immediately change the existing cosmological paradigm, but even strengthened, though for a short time, the belief in its validity. This *classical paradigm* rested on the avowed view that *the universe is eternal and limitless*, that is to say, that the universe has no beginning and no end in time, and has no boundaries in space. These fundamental prerequisites went also hand in hand with the conviction that the universe must be stationary, and not dynamically changing, since otherwise it ought to possess a beginning in time and probably limits in space.

The common embracement of this classical paradigm urged even Einstein to make a correction in his famous tensor equation of general relativity by inserting a special additional Λ-member in it, with the only purpose to save the stationary character of the universe, presented by his fundamental equation.

The later acceptance of a genuine inception of our universe in the *big bang cosmology* was a crucial counter-example to the historically well-established classical paradigm.

One may object that the appearance of big bang cosmology is not simply a rejection of the classical paradigm, but also represents the launch of a new cosmological paradigm, at least since the sixties of the previous century.

My claim is that *the last assumption is not true*. It may be true, of course, concerning the establishment of what I call a *cosmological picture of the universe*, to play the role of a new paradigm. Its relatively wide acceptance is due to the agreement about some features, which characterize our contemporary knowledge about the beginning and the evolutionary change of the universe. These features, however, although being taken as (almost) uncontroversial, do not cover an exhaustive knowledge neither of a complete set of them, nor even display a thorough knowledge of some of their reasons.

Probably (the prevailing amount of) all astronomers and theoretical cosmologists would agree about a list of important features that characterize the universe. It started with a big bang some 13.8 billion years ago. It expands, leaving its early huge inflation aside, with a constant speed for some stage of its evolution, but now its expansion is accelerating. It displays a homogeneous distribution of galactic constellations in a large scale. It has a size of about 92 billion light years.

All such features of the universe stay as important details within the cosmological picture that scientists have drawn of the universe up to nowadays.

This picture, however, can hardly take the place of a full-fledged theory staying at the base of a new cosmological paradigm. It may be the fact that the universe came into existence 13.8 billion years ago, but the real reason about its birth is still unknown. Was the big bang a steady quantum fluctuation (Susskind 2005: chs.3, 11; Hawking, Mlodinov 2005: ch.8; Krauss 2012: ch.9)? Was it a "big crunch" in a reverse collapsing stage of the universe after the end of its expansion that turned into a big bounce? Was the big bang a result of two colliding branes floating in an eleven dimensional spacetime of the so-called M-theory (Greene 2004: ch. IV), or some "play" of space-time geometry, or something else? These are hypotheses upheld by different scientists and demonstrating a *conceptual diversity*.

The material constitution of the hypothetical dark matter, as well as the nature of the dark energy imbedded in spacetime itself, are a subject of controversy among scientists. No cosmologist dares to bet what is the real form of the universe, and whether it is finite or infinite. Contemporary physics has still not reached a "theory of everything," which has the pretension to unify the four fundamental interactions[1] for providing an essential understanding of our universe under the spell of a "God equation," and thus to exculpate Michio Kaku's optimism for this cherished dream (Kaku 2021).[2]

If the answers to the list of all these problems were still not settled, one could hardly insist that we have an established paradigm, in spite of the general contours of the cosmological picture of the universe. The knowledge this picture provides, can enter the conceptual framework of *different* cosmological theories. It is exactly this fact, which demonstrates the lack of an established paradigm. Instead of it, we are witnesses of a whole spectrum of cosmological theories and conceptions. I used the word "spectrum," because the extant range of cosmological theories starts from conceptual systems that still defend amended versions of the ideological core of the classical paradigm about a boundless steady universe,[3] and ends with the multiverse theory, stating the birth and death of myriads of universes, which can exhibit essential differences among each other. This wide and diverge range of theories of the universe cannot certainly constitute an accepted cosmological paradigm.

I raise my claim that contemporary cosmology has went out of its paradigmatic stage not as an intended criticism of Kuhn's well-known methodological view about the paradigm structure of scientific growth. His view has been widely and intensively analyzed (together with its concomitant theses like that of incommensurability) since the first publication of his *The Structure of Scientific Revolutions*. Nevertheless, it was in the Postscript of the second edition of his book, where T. Kuhn presented explicitly the four components of a scientific paradigm, renaming it as "disciplinary matrix.". The first one, pointed as an "important sort of component," he labelled "symbolic generalizations" (Kuhn 1970: 182). They are the formal representations of the theoretical laws. The second component is named "the metaphysical parts of paradigms" (Ibid: 184), which give the conceptual clothing of the theoretical models. Without these first two components, no theory at the center of a disciplinary matrix is possible, and so is the disciplinary matrix itself. If, as I have already shown, there are different cosmological theories, but not a widely accepted one by the scientific community, then the formation of a contemporary cosmological paradigm is failing.

I further suggest that the history of a *complex discipline* like cosmology may go through various historical periods, instead of undergoing one global transition from a pre-paradigmatic towards a paradigmatic stage of its development. While this Kuhn's model implies a quick (and predominantly irrational) change between two consecutive paradigms, the case with cosmology displays a different scenario – a prolonged post-paradigmatic period.

My explanation of this fact rests on the nature of this discipline, which I qualified as a "complex" one. What I mean by this qualification is that contemporary cosmology incorporates in its ontology not only astronomical knowledge, but also physical theories and models about the structure of the microworld, as well as theoretical presentations about the

---

[1]The electromagnetic, the strong nuclear, the weak nuclear, and the gravitational interactions.

[2]There are several suggested models to quantize gravity, but their mechanism of quantization differs from the standard procedure for the quantization of the three remaining force fields by pointing to specific boson particles that transmit their dynamic action. Hence, even if one avowed a success for quantizing gravity, this success cannot yet serve for the cause of a real *unification* of all four physical interactions.

[3]See in this connection Fred Hoyle's steady-state theory https://www.britannica.com/science/steady-state-theory

formation and the further dynamics of stars, galaxies, and clusters of galaxies. All these theoretical models and presentations meet problems of their own, and any essential change in them, or the acquisition of new astrophysical data that must be assimilated by them, affects other aspects of cosmological knowledge. Thus for instance, the recent discovery of the Higgs-boson,[4] or the yet undiscovered nature of the dark matter and the dark energy, have a reflection on our knowledge about the initial states of the universe immediately after the big bang, as well as about its further evolution and future stages.

In addition, I must point here to the quite recent fact that unexpected astrophysical data is now being obtained by contemporary space telescopes working in the range of different radiation specters. This new data cries for an explanation that might require appropriate changes in the lines of the accepted picture of the universe. To this effect, it suffices to mention two such interesting observations. The first one is James Webb's recent discovery of massive galaxies in the early universe, which already existed at only 300-500 million years after the big bang. No doubt, scientists were astonished by this unexpected discovery, since they were certain that there could be no galactic formations at such an early period of the evolution of the universe.[5]

Such discoveries threaten even the validity of what I dubbed the cosmological picture of the universe, not to say of an accepted paradigm, which I claim not to exist. For instance, some voices are now being heard that the universe might be twice older than the 13,8 billion years – the period that has been established to be its real age.

The second interesting observation is on its way to reveal the real nature of the mysterious quasars. Since many years, it was agreed that they are star-like but very bright cosmic objects, exceeding the light energy emitted by ordinary stars. However, the Hubble space telescope "has observed several quasars and found that they all reside at galactic centers. Today most scientists believe that super massive black holes at the galactic centers are the "engines" that power the quasars."[6] The conviction that quasars are not separate star-like objects, but the emitted intensive light radiation by massive black holes, has recently received support by the data obtained from NuSTAR (Nuclear Spectroscopic Telescope Array).[7]

Let me adduce an additional brief comment to my claim that contemporary cosmology is not encompassed by an accepted paradigm. Because an objection may be raised against it. One may yet contend that the contemporary knowledge about the universe incorporated into what I depicted as a received cosmological picture of the universe could pretend to play the role of a paradigm. Scientists who uphold this picture of the universe may differ in some technical details or/and conceptual variations, which do not, however, take them out of the knowledge system of this picture. To this effect, this general cosmological picture is playing the role of a paradigm.

My first argument against this objection is the fact that the diversity of cosmological theories displays *deep conceptual differences*, which surmount the theoretical details about

---

[4] The Higgs boson, known also by the popular name "God particle," "is the fundamental force-carrying particle of the Higgs field, which is responsible for granting other particles their mass" https://www.space.com/higgs-boson-god-particle-explained.

[5] https://phys.org/news/2023-02-webb-massive-galaxies-early-universe.html;
https://www.scientificamerican.com/article/astronomers-grapple-with-jwsts-discovery-of-early-galaxies1/

[6] https://esahubble.org/science/black holes/

[7] https://www.nustar.caltech.edu/news/129

which the proponents of the same paradigm may have disagreements. Since a universe without dark matter or dark energy is quite different from a universe having such essential components. Accepting multiverse theory is a very different understanding from thinking of one single and unique universe, etc. This theoretical diversity cannot be covered by some homogeneous cosmological conception to stay at the center of a paradigm.

As such general conception the received cosmological picture itself could be pointed to. And here comes my second objection. The new observational data, some of which mentioned above, threated even the features of this picture. Cosmologists are no longer unanimous neither about the moment of the birth of the universe, nor even about the veracity of its acceptance. Moreover, even the universal expansion is recently being questioned via a different explanation of the red shift of distant galaxies, the size of the universe becomes a subject of discussion, etc. Hence, the received cosmological picture cannot be declared a paradigm, until its main parameters are not settled so far.

The question "naturally" comes to the fore: "*Why there are many cosmological theories nowadays?*" The answer that could be given after what has been said up to now is that the extant cosmological picture presents only general drawings of the history of the universe, and remains silent about important qualities, which characterize its deep material and dynamic nature. This is a veridical answer, but still not a full one, having in mind the wide spectrum of cosmological theories, containing even theories without an assumed big bang. It appears that there exists a specific explanatory discontent among the upholders of different cosmological theories, although they try to account for the main observational data encompassed by the general cosmological picture. My claim is that this specific discontent is because they stick to different foundational conceptual prerequisites guiding the structure of their theories. These "foundational conceptual prerequisites" can be only of a philosophical and aesthetical nature. Moreover, probably because of this, the upholders of different cosmological theories do not explicitly mention the "ideology" behind their constructions. Notwithstanding this perplexity, I shall start with pointing to the *methodological influence of some philosophical predilections.*

As it is maybe known, the Russian physicist Dmitry Ivanenko has asked six noble prizewinners who were his guests in Moscow State University "Lomonossov" to write on the wall of his room curt maxims they estimated to be a kind of conceptual guidelines.[8] Prigogine's sentence is this: "*Time precedes existence.*" This sentence is not a catchword, but encodes an important aspect of Prigogine's ontological view about the time and the universe, turned to be a philosophical belief for him. It is embedded in his conviction that our universe has a beginning and probably an end, but time is more fundamental than the existence of a separate universe, since it has no beginning and no end within the limits of a universe. The big bang has really happened in time, but time did not start with the birth of the universe, it is a cosmic essence. Prigogine's methodological guideline is his philosophical predilection that our universe had a beginning – it started by some kind of a quantum phase transition from a preuniverse, as he used to call the quantum vacuum – *but it had a beginning in time* (Prigogine 1996: ch.VIII). Time precedes the existence of the universe, because it is meta-universal.

---

[8]Among them is the famous Niels Bohr's dictum: "contraria non contradictoria sed complementa sunt."

There is another philosophical predilection, which also ascribes an eternity to time, that is to say, assumes that time has no beginning and no end. However, it does not expel time out of our universe, as Prigogine does. In this way, this philosophical predilection presupposes the eternity of the universe *per se*, in the sense that it has no beginning, and will have no end. Such is for instance Fred Hoyle's steady-state theory that was pointed to in fn. 3. In order to keep its philosophical prerequisite intact, this cosmological theory attracts the assumption that the observed expansion of the universe does not lead to its not stationary character, since the material density of the universe is kept constant, because of a permanent creation of new matter.

Accepting a big bang cosmology does not necessarily mean that our universe started from a singularity, in which all its matter/energy was stuck, and all the known physical laws cease to be valid. Such is the conceptual prerequisite of Stephen Hawking. Instead of such a singularity, he prefers to postulate his so called "no boundary" idea. For this purpose, he introduces an imaginary time, which resembles the three spatial dimensions. Thus spacetime becomes homogeneous, finite, and without boundaries. I shall not present details of his cosmological construction. It is here more important to point to Hawking's avowal that his idea is not deduced from some principle, but is "put forward for aesthetic or metaphysical reasons" (Hawking 1989: 144). This is something I wanted to stress by my claim that philosophical predilections, though staying at the heuristical background of a cosmological theory, may affect its ontological construction.

Aesthetical predilections, often intertwined with philosophical ones, can also guide, if I can say so, the ontological spirit of cosmological theories. I mean here that sometimes scientists mention a requirement for beauty and elegance of a scientific theory.

However, which elements of a theory this requirement could concern? Let me turn back to the already discussed Prigogine's maxim that time precedes existence. Above it, on the wall of prof. Ivanenko's room stands another one that belongs to the famous physicist Paul Dirac. His maxim states: "*Physical law should have mathematical beauty.*"

Mathematical beauty can thus be referred to the formal expression of the laws of a mathematized theory, and such are all of the quantum theoretical schemes at the base of contemporary cosmological theories. As for the elegance, it can concern the number of principles and laws of a theory, and/or a given theory as a whole. However, what I would like to explicate here is not what it means a physical law to possess mathematical beauty, or why a theory may be estimated as an elegant one. Even more that scientists themselves do not have shared aesthetical criteria when applying these qualifications, but show disagreements about them. What I am arguing here for is the claim that aesthetical prerequisites can affect the formation of cosmological theories.

My further specification is that scientists usually stick to aesthetical assessments not for a cosmological theory as a whole, but for its underlying quantum physical theories. String theory is one of them. As Roger Penrose clearly explains, string theory pretends to depict an ontological situation deep in the micro-world at a Planck scale, which could be verified by high-energy experiments unreachable by contemporary accelerators. For this reason, lack of experimental support requires that "additional criteria of *mathematical elegance* and simplicity must be invoked" (Penrose 2016: 3). However, in the world of theoretical physics, Penrose specifies, aesthetical judgements are often contaminated by "elements of fashion," and thus cannot be relied on as an objective criterion (Ibid: 4).

Indeed, scientists who estimate string theory to be an elegant one are inclined to advertise a contemporary cosmological conception known as a multiverse theory. It accepts the existence of an enormous number of universes, besides our own, which are characterized by a different constitution of their material structure and physical laws determined by the internal features of the tenth-dimensional space of the string theory.

On the contrary, there are scientists who do not find string theory to be elegant and criticize it for different reasons. Among them (though not always active critics of the string theory) are adherents of the so-called standard model about the nature of the elementary particles and their basic interactions, who go on to elaborate the received inflationary view of the big bang cosmology.

By these reflections I tried to show that philosophical and aesthetical predilections, although set in a background worldview, can exert an influence on the formation of cosmological theories.

References

Greene, Brian. 2004. *The Fabric of the Cosmos: Space, Time, and the Texture of Reality.* Alfred Knopf.

Hawking, Stephen W. 1989. *A Brief History of Time. From the Big Bang to Black Holes.* Bantam Books.

Hawking, Stephen with Leonard Mlodinov. 2005. *A Briefer History of Time.* Bantam Press.

Kaku, Michio. 2021. *The God Equation: The Quest for a Theory of Everything.* New York: Doubleday.

Krauss, Lawrence M. 2012. *A Universe from Nothing. Why There Is Something Rather than Nothing.* New York, London, Toronto, Sidney, New Delhi: ATRIA Paperback.

Kuhn, Thomas S. 1970. *The Structure of Scientific Revolutions.* Chicago: The University of Chicago Press.

Penrose, Roger. 2016. *Fashion, Faith, and Fantasy in the New Physics of the Universe.* Princeton and Oxford: Princeton University Press.

Prigogine, Ilya. 1996. *La fin des certitudes. Temps, chaos et les lois de la nature.* Paris: Odile Jacob.

Susskind, Leonard. 2006. *The Cosmic Landscape.* New York, Boston, London: Back Bay Books.

Part III

ONTOLOGY OF TIME

Kyley Ewing (Ed.), *Spacetime Conference 2024. Selected peer-reviewed papers presented at the Seventh International Conference on the Nature and Ontology of Spacetime, 16 - 19 September 2024, Albena, Bulgaria* (Minkowski Institute Press, Montreal 2025). ISBN 978-1-998902-44-6 (softcover), ISBN 978-1-998902-45-3 (ebook).

7 MatterSpaceTime (MST) Theory

Prakash Bhat

Abstract This paper presents the MatterSpaceTime (MST) theory, which integrates matter and spacetime into a single mixed-dimensional fabric. By extending spacetime to include regions of varying dimensionality influenced by localized energy concentrations, MST theory provides a novel framework for understanding fundamental physical phenomena. The theory suggests that extra dimensions emerge dynamically where matter is present, eliminating the need for compactification in string theory. MST offers new insights into quantum gravity, entropy, etc. This paper delves into the mathematical underpinnings of MST, explores its philosophical implications, and discusses applications in quantum mechanics interpretations, cosmology and machine learning.

1 Introduction

Major advances in theoretical physics have come from redefining frame of reference, space and time. Cartesian coordinates enabled development of calculus and classical physics. Integration of space and time into spacetime eases the explanation of various special relativity phenomena. Curvature of spacetime ushered in General Relativity and Big Bang Cosmology.

A logical question: why stop at the "curvature of spacetime fabric"? If spacetime can curve, can the spacetime fabric have lumps(or bumps, blisters or bubbles; whichever suits your imagination; This is depicted in the figure 1)? Do these lumps require a locality(region, area, piece or part) of spacetime at a higher dimensionality than the surrounding? Can these lumps be considered as matter which is localized energy? These intuitions suggest that matter can be integrated with spacetime. Such integrated mixed dimensional fabric is called MatterSpaceTime(abbreviated as MST) in this document. This helps to intuitively understand important unresolved questions in physics and philosophy. Such explanations are collated as MatterSpaceTime(MST) theory.

String theory needs the existence of extra dimensional spaces in its search for a theory of everything. According to String theory we don't see extra dimensions because they are compactified. MatterSpaceTime helps us to visualize extra dimensions as locally created dimensions without the need of compactification. According to MatterSpaceTime theory, extra dimensions are created only in the area where matter particles in spacetime are present.

MatterSpaceTime theory is background-independent like loop quantum gravity. MatterSpaceTime theory could be a general theory reducing to string theory for regions of higher dimensionality and to loop quantum gravity for regions with low dimensionality without matter. Various innovations of MST such as mixed dimensions are compatible with both discrete and continuous spacetime and are relevant to most of the quantum gravity theories and may help to overcome some of the obstacles they face.

Kyley Ewing (Ed.), *Spacetime Conference 2024. Selected peer-reviewed papers presented at the Seventh International Conference on the Nature and Ontology of Spacetime, 16 - 19 September 2024, Albena, Bulgaria* (Minkowski Institute Press, Montreal 2025). ISBN 978-1-998902-44-6 (softcover), ISBN 978-1-998902-45-3 (ebook).

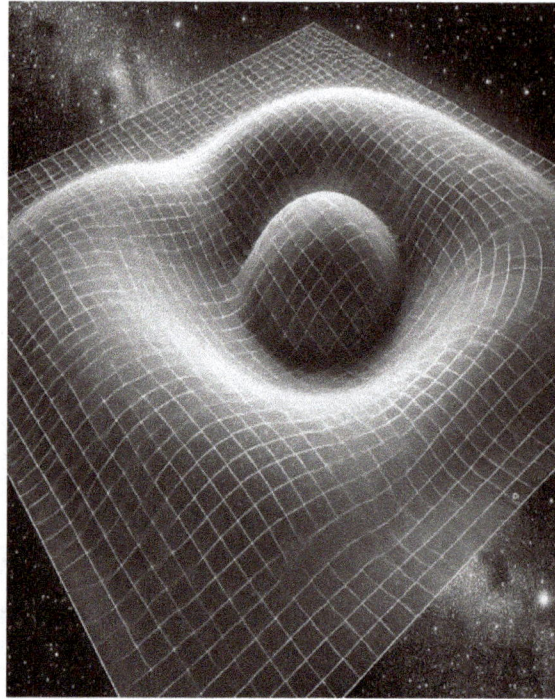

Figure 1: Mixed dimensional lumps

As energy concentrates together in a small area, spacetime lumps appear resulting in matter particles. Extra dimensions are created in the lumpy areas as these quantum particles are getting created from the localized energy. In General Relativity curvature can vary from place to place(or point to point). Similarly in MatterSpaceTime Theory, dimensionality can vary across SpaceTime fabric. So instead of "matter being in SpaceTime" as in String Theory/General Relativity, we have matter as an integral part of the MatterSpaceTime; This view is more unifying.

We have various dualities; In physics, Gravity and Quantum Mechanics need theories of quantum gravity. In philosophy, materialism and idealism are conflicting viewpoints. MatterSpaceTime can be the starting point to unify these dualities. It is interesting to see how these dualities can emerge from the natural division of MatterSpaceTime.

2 Mathematical Framework for Mixed Dimensions

This section discusses how various branches of mathematics can be extended to incorporate mixed dimensions(different parts of an entity having different number of dimensions; multiple dimensions) if they don't already include them.

2.1 Non-Euclidean Geometry

The development of alternative parallel postulates has led to the exploration of geometries beyond the traditional Euclidean framework, resulting in Non-Euclidean geometries. In

MST theory, these geometries are further generalized to accommodate regions with different numbers of dimensions, creating a framework known as mixed-dimensional geometry.

Mixed-Dimensional Geometry

In MST, different regions of the geometry can have different numbers of dimensions. This extension allows for a more flexible and comprehensive description of spacetime, particularly in the presence of localized matter concentrations or "lumps" that can alter the local dimensionality.

- **Local Dimensionality**: Each region R_i of the space M has a specific dimensionality d_i, such that $M = \bigcup_i R_i$, where d_i can vary.

- **Dimensional Function**: A dimensional function $\mathrm{Dim}(x)$ assigns a dimension d_i to each point $x \in R_i$.

- **Generalization of Euclidean Postulates**: Mixed-dimensional geometries conform to Euclidean postulates (excluding the parallel postulate), similar to non-Euclidean geometries, but allow for the local variation in dimensionality.

Mathematical Representation

Consider a space M where the dimensionality varies:

$$M = \bigcup_i R_i \quad \text{with} \quad \mathrm{Dim}(x) = d_i \quad \text{for} \quad x \in R_i.$$

This framework enables the modeling of complex geometrical structures where the local properties of space can change dynamically, influenced by the presence of matter and energy.

2.2 Manifolds in Mixed Dimensions

Manifolds are fundamental objects in geometry and topology that generalize the notion of curves and surfaces to higher dimensions. In MST, we extend the concept of manifolds to accommodate mixed dimensions.

Definition of Mixed-Dimensional Manifolds

A mixed-dimensional manifold is a topological space that is locally homeomorphic to Euclidean spaces of varying dimensions. For each point x in the manifold M, there exists a neighborhood U_x that is homeomorphic to an open subset of $\mathbb{R}^{d(x)}$, where $d(x)$ is the local dimension at x.

Charts and Atlases

An atlas for a mixed-dimensional manifold M consists of a collection of charts $\{(U_\alpha, \phi_\alpha)\}$ such that $U_\alpha \subset M$ and $\phi_\alpha : U_\alpha \to \mathbb{R}^{d(x)}$ for $x \in U_\alpha$.

Algebraic Topology

Algebraic Topology structures such as Simplicial(generalization of triangle to n-dimensions) Complexes and CW Complexes as shown in figure 2 already utilize mixed dimensions. Spin foam theories (reference [8]) of Loop Quantum Gravity involve 2-complexes and could be natural places where application of mixed dimensions can be further explored.

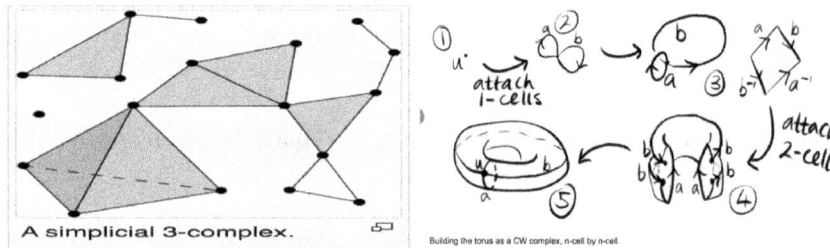

Figure 2: Simplical and CW complexes

2.3 Challenges of Mixed Dimensions

While mixed-dimensional geometry provides a powerful framework for MST, it introduces several challenges that need to be addressed to fully realize its potential.

Consistency and Continuity

Ensuring consistency and continuity across regions with different dimensions is a significant challenge. The transition between regions R_i and R_j with different dimensionalities d_i and d_j must be handled carefully to maintain the overall coherence of the space.

Mathematical Tools and Techniques

Developing appropriate mathematical tools and techniques to handle mixed-dimensional spaces is crucial. This includes:

- **Differential Calculus**: Extending differential calculus to accommodate functions and fields defined over regions with varying dimensions.

- **Integration Theory**: Formulating integration over mixed-dimensional manifolds to ensure proper measure and integration techniques.

- **Tensor Calculus**: Adapting tensor calculus to operate in mixed-dimensional contexts, ensuring that tensors are well-defined across regions with different dimensionalities.

Physical Interpretation

Understanding the physical implications of mixed-dimensional geometry is essential. This involves:

- **Impact on Physical Laws**: Investigating how the variation in dimensionality affects fundamental physical laws, such as those governing gravity, electromagnetism, and quantum mechanics.

- **Experimental Predictions**: Deriving testable predictions from the theory to validate the existence and effects of mixed-dimensional regions in spacetime.

Computational Challenges

Modeling and simulating mixed-dimensional spaces pose significant computational challenges. Developing efficient algorithms and numerical methods to handle the complexity of such spaces is crucial for practical applications.

2.4 Applications and Implications

The concept of mixed-dimensional geometry in MST has far-reaching applications and implications across various fields of physics and mathematics.

Cosmology and Astrophysics

Mixed-dimensional geometry can provide new insights into the structure of the universe, particularly in regions with extreme gravitational fields, such as near black holes or in the early universe during the Big Bang.

Quantum Gravity and Field Theory

Incorporating mixed dimensions into quantum gravity and field theory frameworks can lead to a deeper understanding of the fundamental nature of spacetime and its interactions with matter at quantum scales.

Mathematical Extensions

The study of mixed-dimensional geometry can lead to new mathematical discoveries, extending the current understanding of differential geometry, topology, algebraic structures, and category theory.

Computer Science : Data and AI/ML(Artifical Intelligence/Machine Learning)

NoSQL databases, JSON (JavaScript Object Notation) & XML(Extensible Markup Language) data formats already allow representing mixed dimensional values. AI/ML algorithms optimally utilizing mixed dimensions need to be developed which can save CPU/GPU computation time needed for ML training and inference.

2.5 Conclusion

The extension of non-Euclidean geometry to mixed-dimensional spaces in MST theory offers a rich and versatile framework for exploring the complexities of spacetime. While it presents several challenges, it also opens up numerous possibilities for advancing theoretical physics and mathematics. Addressing these challenges through the development of new mathematical tools, physical interpretations, and computational methods will be key to unlocking the full potential of this approach.

3 The Ontology of MatterSpaceTime: Dimensions, Universe, and Reality

The ontology of MatterSpaceTime (MST) involves a profound exploration of dimensions, the nature of the universe, and the entities and relationships within it. By simplifying the ontological framework to a single unified entity—MatterSpaceTime—this chapter explores how dimensions emerge, the role of the universe as an all inclusive system, and the implications for physics and philosophy.

3.1 The Nature and Ontology of Dimensions : Traditional perspectives

Dimensions are fundamental constructs in both mathematics and physics, serving as tools to describe the universe's structure and behavior. This section examines their nature, roles, and philosophical underpinnings.

- **Mathematical Construct:** Dimensions are directions in which movement or measurement can occur. In Euclidean geometry, they correspond to independent axes (e.g., x, y, z in 3D space), forming the basis for spatial representation and calculation.

- **Conceptual Frameworks Across Theoretical Physics:**

    - **Classical Mechanics:** Relies on three spatial dimensions and one temporal dimension.

    - **Relativity:** Specialy Relativity merges space and time into a four-dimensional spacetime continuum; In General Realtivity spacetime is locally flat but globally curved with different curvatures based on "metric".

    - **Quantum Mechanics:** Implicitly incorporates two dimensions (real & imaginary components of wavefunctions) for each coordinate(x,y,z) for each point leading to infinite dimensional Hilbert Space.

    - **Loop Quantum Gravity:** Space and Time are not fundamental and are discrete entities at the most fundamental level.

    - **String Theory and Beyond:** Proposes additional dimensions, often 10 or 11, to unify fundamental forces and achieve mathematical consistency.

- **Philosophical Perspectives:** Dimensions may be fundamental properties of the universe, existing independently of observation, or they could be emergent phenomena arising from more basic entities. Dimensions form integral components of the universe's structure, shaping physical laws and theoretical models. Philosophers have debated whether space is a substance (substantialism) or a relational construct (relationism). Relativity introduced the block universe model, which views time as a dimension akin to space. Structural realism further emphasizes metric structures over spatial points, offering a more abstract perspective.

3.2   MatterSpaceTime Perspective

In MST theory, nature, structure and number dimensions are not fixed or external; Instead, they emerge locally as intrinsic patterns within the universe. This perspective shifts the focus from predefined dimensionality to dynamic, context-dependent structures.

The Universe as all inclusive system

If the Universe is defined as encompassing all that exists—concepts, logic, mathematical forms, patterns, and physical laws—then nothing can be external/extrinsic to it. This all inclusive definition implies that the number and nature of dimensions in any region are emergent properties of the universe itself. Dimensions are not imposed externally but arise as patterns within the system. MST simplifies ontology by positing qualitatively and quantitatively single fundamental ontological entity : MatterSpaceTime. All the observed properties and phenomena of the Universe need be derived as emergent.

- **Unified Framework:** MST reconciles the dichotomy between "vacuum/spacetime" and "something/matter." In regions of lower dimensionality, MatterSpaceTime manifests as spacetime, while in higher dimensionality, it emerges as matter.

- **Occam's Razor:** This parsimony adheres to Occam's Razor, positing that no additional fundamental ontological entities are needed to explain observable emerged phenomena. If one tries to ontologically deny existence of even this single entity(MST), then we have to attribute properties to "nothing" to derive observed phenomena which promotes "nothing" to a commonsense "something" defeating the purpose of denial of even a single entity. Hence we can start with single ontological entity MST and see how see can derive emergence of observed phenomena.

Quantum Nature and Self-Division

The quantum nature of reality emerges from the self-division of MatterSpaceTime. Max Planck's derivation of his radiation law underscored this division, introducing Planck's constant h and the concept of quantization.

- **Quantum nature emerging from continuum:** Quantum mechanics reveals a discreteness in reality, with Planck's constant representing the irreducible quantum of action. But this discreteness emerges from the wave function defined on the continuum. This shows how you can have discreteness arising out of a single continuous entity. This is like the multiplicity of fingers emerging from our single hand.

- **Duality of MatterSpaceTime Division: Whole or Divided?** Just like Wave/Particle duality, there are contexts in which considering MatterSpaceTime or Universe as a whole make sense; there are situations where MatterSpaceTime is considered to have divided itself into parts, components, regions etc. There might be situations where both whole and divided perspective might be needed: for example finger in a hand can be considered divided at the same time they are joined at the hand. When considered divided it can be integer, real or using any of the normed division algebra where division has meaning.

- **Algebra of Division:** Division algebras provide a mathematical foundation for understanding how MatterSpaceTime divides itself into quantum components, connecting the discreteness of quantum mechanics to the broader ontological structure of MST. Hence Division Algebra is discussed in the next section.

3.3 Conclusion

The ontology of MatterSpaceTime offers a unified framework for understanding dimensions, quantum mechanics, and the structure of the universe. By reducing reality to a single entity with emergent properties, MST simplifies complex ontological debates while preserving explanatory power. This approach aligns with both philosophical principles, such as Occam's Razor, and the empirical findings of modern physics. Dimensions, rather than being fixed constructs, emerge dynamically within the MST framework, reconciling classical and quantum views. This perspective invites further exploration of the interplay between mathematics, physics, and philosophy in elucidating the nature of reality.

4 Division Algebras, Cayley-Dickson Construction, and Their Relevance to Mathematical Physics

4.1 Introduction

Division algebras and the Cayley-Dickson construction play a crucial role in mathematical physics, providing a framework for understanding various physical phenomena. These structures are essential for explaining why complex numbers are required in quantum mechanics, how quaternions simplify Maxwell's equations, and the ongoing research using octonions to understand the Standard Model (SM) of particle physics.

4.2 Division Algebras

Definition and Properties

A division algebra is an algebra over a field where division is possible, except by zero. The four normed division algebras, along with their unique properties, are:

- **Real Numbers** ($\mathbb{R}$): Commutative and associative.

- **Complex Numbers** ($\mathbb{C}$): Commutative but not associative.

- **Quaternions** ($\mathbb{H}$): Neither commutative nor associative but still a division algebra.

- **Octonions** ($\mathbb{O}$): Neither commutative nor associative but satisfy a weaker form of associativity called alternativity.

Relevance to Quantum Mechanics and Electromagnetism

- **Complex Numbers in Quantum Mechanics**: Complex numbers are essential in quantum mechanics due to their role in describing wave functions and probability amplitudes. The Schrödinger equation, which governs the behavior of quantum systems, inherently relies on complex numbers.

- **Quaternions and Maxwell's Equations**: Maxwell's equations, which describe the behavior of electromagnetic fields, can be elegantly formulated using quaternions. This reduces the set of four equations to a single quaternionic equation, simplifying the mathematical framework and highlighting the underlying symmetry of electromagnetism.

- **Octonions and the Standard Model**: Ongoing research explores using octonions to understand the Standard Model of particle physics. Octonions, with their higher-dimensional structure, offer a potential framework for unifying different fundamental forces and particles, providing insights into the deeper structure of physical laws.

4.3   Cayley-Dickson Construction

Construction Process

The Cayley-Dickson construction is a recursive method to generate a new algebra from an existing one, effectively doubling the number of dimensions at each step. Starting with the real numbers ($\mathbb{R}$), the construction proceeds as follows:

- **Complex Numbers** ($\mathbb{C}$): Constructed by defining $a + bi$ where $i^2 = -1$.

- **Quaternions** ($\mathbb{H}$): Constructed by defining pairs of complex numbers $(a+bi)+(c+di)j$ where $j^2 = -1$ and $ij = -ji$.

- **Octonions** ($\mathbb{O}$): Constructed by defining pairs of quaternions $(a+bi+cj+dk)+(e+fi+gj+hk)l$ where $l^2 = -1$ and follows specific multiplication rules.

Product Definition

For algebras generated by the Cayley-Dickson construction, the product of two elements (a, b) and (c, d) is defined as:

$$(a, b) \cdot (c, d) = (ac - d^*b, da + bc^*)$$

where a, b, c, d are elements of the original algebra, and d^* and c^* denote the involution (conjugate) of d and c, respectively.

Involution

An important aspect of the Cayley-Dickson construction is the conjugation, which is a type of involution operation. For an element $x = (a, b)$ in the algebra generated from an existing algebra, the involution is defined as: $x^* = (a^*, -b)$ where a^* is the conjugate in the original algebra.

Norm

The norm of an element $x = (a, b)$ in an algebra generated by the Cayley-Dickson construction is defined as:

$\|x\| = \sqrt{xx^*} = \sqrt{aa^* + bb^*}$ where a, b are elements of the original algebra and a^*, b^* are their involutions.

Power-Associativity

The Cayley-Dickson construction generalizes normed division algebras to power-associative algebras. An algebra is power-associative if it satisfies the associative property for powers of any element. This means that for any element x in the algebra and for any integers n, m, the powers of x satisfy the associative property:

$$(x^n x^m)x = x^n(x^m x)$$

for all n, m. This property is weaker than full associativity but is sufficient to maintain a consistent algebraic structure across higher dimensions.

4.4   Implications in Mathematical Physics

Unified Frameworks

Exploring the Cayley-Dickson construction provides a unified framework for understanding:

- Quantum Mechanics: The necessity of complex numbers.

- Electromagnetism: The compact expression of Maxwell's equations using quaternions.

- Particle Physics: The potential unification of the Standard Model using octonions.

4.5 Conclusion

Division algebras and the Cayley-Dickson construction offer deep insights into the structure and behavior of quantum systems. By understanding the mathematical foundation and implications of these algebras, researchers can explore unified theories and coherent frameworks in mathematical physics.

5 Quantum Mechanics in MatterSpaceTime (MST) Theory

Quantum mechanics (QM) in MatterSpaceTime (MST) theory provides a unique perspective on the fundamental nature of quantum systems, integrating ideas from quantum theory, classical physics, and high-dimensional algebraic structures. This section elaborates on the key ideas of quantum mechanics within MST theory, focusing on the emergence of classical physics, the independence of quantum systems, wavefunction collapse, superposition, and the role of high-dimensional structures in quantum phenomena.

5.1 Quantum Systems as Components of MatterSpaceTime

In MST theory, each smallest component of MatterSpaceTime is a quantum system that possesses all the properties from which classical physics, logic, and mathematical patterns emerge. This analogy likens each quantum system to a biological cell, with properties that determine the behavior of larger structures.

Emergence of Classical Physics from Quantum Systems

Just as DNA encodes the information necessary for the development of an organism, each quantum system encodes the fundamental properties from which classical physics emerges. The interaction of quantum systems leads to the formation of macroscopic classical objects, governed by classical physics.

- **Sociology of Quantum Systems:** The behavior of a quantum system is influenced by the freedom and interactions of other quantum systems in the universe. This "sociology" of quantum systems gives rise to the emergent laws of physics and mathematics.

- **Emergence of Classical Laws:** Classical laws such as Newton's laws of motion, Maxwell's equations, thermodynamics and special & general relativity can be viewed as emergent properties resulting from the interactions of a vast number of quantum systems. The collective behavior of these non-deterministic systems gives rise to deterministic classical phenomena.

- **Equations:** The emergence of classical laws from quantum systems can be described mathematically using statistical mechanics and decoherence theory. For example, the density matrix $\rho(t)$ of a quantum system evolves over time as:

$$\rho(t) = \sum_i p_i |\psi_i\rangle \langle \psi_i|$$

where p_i are the probabilities of different states $|\psi_i\rangle$. As interactions between quantum systems increase, the off-diagonal elements of $\rho(t)$ tend to zero, leading to classical behavior.

5.2 Quantum Systems in the Block Universe

In MST theory, each quantum system has its own internal "clock," but until an interaction, observation, or measurement occurs, the quantum system moves along in a streamlined flow with the "cosmic time" of the block universe. This synchronization with the block universe's time gives the appearance of unitary evolution, where the quantum system follows the spacetime flow. However, when a measurement or interaction occurs, the quantum system's dimensionality increases leading to the independent parameters other than block universe's relativistic time. These additional independent dimensions/parameters enable the probabilistic nature of wavefunction collapse and the superluminal nature of quantum entanglement correlations.

Streamline Comsic River Flow in the Block Universe

The evolution of quantum systems in the block universe can be likened to the flow of a river. While each quantum system has its own internal clock, it moves in a streamlined flow along the cosmic river as if there is a cosmic time. This results in unitary evolution, where the quantum system follows the smooth flow of the block universe's time, just as water particles in a river follow a continuous streamline. Cosmic time is an emergent property and not fundamental absolute time. When parts of the universe moving along in a streamline flow their clocks are synchronized and it looks as if they all following one clock: cosmic time. But the cosmological parts bifurcates into two streams, then there will be two cosmic clocks. Hence changes in the parts of the universe lead to emergence of cosmic clocks based on their streamlines/synchronized flow of changes.

Turbulence, Wavefunction Collapse and Randomness

Measurement, interaction, or observation introduces "turbulence" into the smooth flow of the cosmic river. This turbulence, caused by parts of the river themselves (the quantum systems and their interactions), leads to the quantum system's dimensionality increase. The quantum system evolves utilizing these newly available internal additional dimensions/parameters/internal clock independent of the environment's cosmic clock. This is like parts of the river disengaging from the flow of a river in turbulence. This leads to the wavefunction collapse and the appearance of probabilistic outcomes with respect to environment's cosmic clock.

Superluminal Correlations and Entanglement

The dimensionality increase of quantum systems from the block universe also explains the superluminal correlations observed in quantum entanglement. When particles are brought together to be entangled an extra dimension private channel is created between them with zero metric distance. In this channel, distance is zero even though they are moved far along in the environmental/cosmic metric distance. When entangled particles are measured, they can exchange correlations using this zero metric distance channel. This leads to correlations to manifest instantaneously across vast distances. This is how the Superluminal Correlations observed in entangles particles like Bell Pairs. It is important to note that only correlations, not information, are exchanged between entangled particles. After measurement, the particles continue to move with the cosmic spacetime river, maintaining their alignment with the universe's overall flow. This resolves EPR Paradox and Einstein's concern about "spooky action at a distance."

Implications for Quantum Mechanics and Spacetime

The existence of higher-dimensional spaces in which entangled wavefunctions exchange correlations and evolve suggests a deep connection between quantum mechanics and the structure of spacetime. MST theory provides a framework in which the non-locality observed in quantum entanglement is a natural consequence of the underlying higher-dimensional structure with zero metric distance.

This perspective offers new insights into the nature of quantum correlations and their relationship with the geometry of spacetime. It also opens up possibilities for exploring how other quantum phenomena might be understood in terms of higher-dimensional interactions, potentially leading to new predictions and applications in both quantum mechanics and cosmology.

5.3   Related Quantum Gravity and Unification Theories

Loop Quantum Gravity

Loop Quantum Gravity (LQG) [8] is a non-perturbative, background-independent approach to quantizing spacetime. It reformulates general relativity using Ashtekar variables, in which the gravitational field is expressed as an SU(2) gauge field rather than a metric. These variables enable one to define loop variables (holonomies): the Wilson loops that track how this gauge connection changes around closed paths. Such loops provide the basis for the so-called spin network states, where edges carry spin representations of SU(2), and nodes encode how these spins intertwine. Over time, spin networks evolve into spin foams, offering a covariant "path integral" picture. LQG's major achievements include showing that geometric operators for area and volume have discrete spectra at the Planck scale, yielding a natural candidate for quantum discreteness of spacetime. It has also provided insights into black hole entropy by counting microscopic states of spin networks intersecting a horizon. Open challenges include clarifying the semiclassical limit (recovering smooth spacetime at large scales), coupling matter fields comprehensively, and making precise experimental predictions.

Twistor Theory

Twistor Theory [4], pioneered by Roger Penrose, is a geometric framework that aims to unify quantum mechanics and gravity through a fundamentally different representation of spacetime. In Twistor space, points in ordinary 4D spacetime correspond to certain algebraic conditions (projective lines) within a higher-dimensional complex manifold. By reformulating fields and interactions in terms of twistor variables rather than spacetime coordinates, the theory aspires to simplify the description of gravitational and quantum fields and possibly avoid certain divergences. Twistor Theory has found success in certain calculations—particularly in scattering amplitudes for gauge theories—by exploiting conformal symmetry and complex analytic methods. However, it remains an ongoing research area to fully integrate twistors into a complete theory of quantum gravity and to match all aspects of our observed universe. Despite these challenges, Twistor Theory continues to inspire new insights in mathematical physics and could ultimately provide an elegant path toward a background-independent approach to unification.

String Theory

String Theory [10] (and its extensions like M-theory) proposes that fundamental one-dimensional strings, rather than point particles, underlie all matter and interactions, including gravity. In its simplest form, each vibrational mode of a string corresponds to a different particle species, unifying gauge bosons with gravitons in a single framework. String Theory naturally introduces extra spatial dimensions (commonly 6 or 7 beyond our observed 3+1) which are typically assumed to be compactified on tiny scales (e.g., Calabi–Yau manifolds). Over the years, String Theory has yielded remarkable consistency checks, such as anomalies canceling out under certain conditions, and dualities like T-duality and S-duality that unify different string pictures into a single coherent theory. It also inspired the groundbreaking AdS/CFT correspondence, linking gravity in higher-dimensional anti-de Sitter space to conformal field theories in lower dimensions. However, substantial challenges remain in deriving the exact low-energy spectrum matching our Standard Model, fixing the form of these extra-dimensional compactifications, and pinning down testable predictions in accessible energy regimes. In addition, a notable challenge is that supersymmetry (SUSY)—a crucial symmetry in many string-derived models—has not been found experimentally, even at high-energy colliders such as the LHC. Other open issues include specifying the precise compactification scheme (to reproduce the Standard Model) and extracting testable predictions at currently achievable energies.

Oppenheim's Post-Quantum Theory of Classical Gravity

Jonathan Oppenheim and collaborators have developed a post-quantum theory of classical gravity framework [11] (sometimes also referred to as "gravitational post-quantum theories" or "quantum-classical hybrids") to tackle the tricky interplay between quantum mechanics and classical general relativity. The core idea is to propose that gravity itself might remain classical—governed by a classical metric—while matter is fully quantum. This approach aims to understand whether it is consistent to couple quantum matter to a classical gravitational field without introducing inconsistencies or violating the no-go theorems usually invoked in semi-classical gravity. Key aspects involve investigating how quantum coherence, decoherence, and entanglement behave under a classical metric field and whether such setups can recover conventional predictions (e.g., the Einstein field equations) in appropriate limits.

Although these hybrid ideas could potentially circumvent certain conceptual obstacles of fully quantizing gravity, they face major challenges in ensuring internal consistency, preserving causal structure, and explaining black hole thermodynamics and information flow in strongly curved spacetimes.

5.4 Quantum Gravity in MatterSpaceTime (MST) Theory

MatterSpaceTime (MST) theory resolves the longstanding tension between general relativity and quantum mechanics by introducing the concept of variable dimensionality. In this framework, the pseudo-Riemannian manifold of general relativity is generalized to a hypercomplex manifold with dynamic mixed dimensionality. This reinterpretation eliminates the fixed number of dimensions which are the vestiges of "absolute time and space," offering a unified perspective that integrates quantum phenomena, such as wavefunction collapse and entanglement, with spacetime geometry.

Special relativity eliminated separate absolute space and time but introduced "spacetime" by adding observer velocity as a dependency. General relativity further extended this framework by introducing locality being observed as a dependence, thereby embedding possibly separate tangent spacetime at each point on an observer dependent pseudo-Riemannian manifold. This manifold and the matter/energy within it "mutually" controls one another as specified in Einstein's Equations of General Relativity.

General relativity relies on the key assumption that local spacetimes align smoothly along a differentiable pseudo-Riemannian manifold. A natural question arises: who or what enforces this smooth alignment? This assumption, reminiscent of classical notions of "absolute time and space," may represent the last relic of absolutism in modern physics.

MST challenges this paradigm by allowing spacetime itself to exhibit variable dimensionality because of which a quantum systems spacetime could create additional dimensionality during measurement and enganglement. When the GR constraint that local spacetimes should smoothly vary is relaxed, the peculiarities of quantum mechanics—such as random wavefunction collapse and non-local entanglement—are no longer anomalies but intuitive consequences of the dynamic turbulent manifold structure. Variable dimensionality provides an explanatory framework for these quantum behaviors by linking them to the turbulent geometry of spacetime arising out of variable dimensionality. This turbulent manifold is akin to Loop Quantum Gravity Space Foam but shows variable dimensionality.

By dropping the assumption of a strictly differentiable pseudo-Riemannian manifold, MST achieves the hard problem of integrating quantum mechanics and general relativity. The smooth, differentiable spacetime of general relativity becomes an emergent behavior of a manifold with variable dimensionality, aligning seamlessly with the probabilistic and non-local nature of quantum systems. This shift removes the last vestiges of "absolute time and space," replacing them with a dynamic, context-dependent variable dimensional spacetime that unifies quantum and relativistic physics under a single framework.

Quantum theory is fundamentally linear, while gravity exhibits non-linear characteristics. This naturally raises the question of how a non-linear system might emerge from a linear framework. When numerous quantum systems interact, their collective behavior could give rise to non-linear structures, such as the curvature of spacetime.

These gravity non-linear structures, in turn, exhibit a tendency to preserve their "form," leading to emergent phenomena like gravity influencing the behavior of the underlying quantum systems. This is like a whirlpool in a river using its energy to preserve itself. This type of upper level reality forms interaction with lower layer objects could result in gravity-induced decoherence, as proposed by Penrose.

Quantum linearity can be explained by saying they don't have any parts to host the non-linearity. Just like universe does not have anything external to it, Quantum Subsystem does not have anything internal underneath its quantum nature demonstrating its most fundamental nature.

Black hole singularity of general relativity is avoided in MST by the phenomenon of Pseudo-Rimeanninan manifold yields to increasingly higher dimensionality from event horizon and to the center of blackhole to accommodate huge amount of matter/energy.

MST and Loop Quantum Gravity

Loop Quantum Gravity (LQG) centers on a fundamentally discrete geometry encoded by spin networks, with spacetime emerging from the combinatorial structure of SU(2) holonomies, whereas MST posits a continuous underlying fabric whose local dimensionality changes in the presence of matter-energy lumps. Despite these formal differences, both frameworks share the principle of background independence, treating geometry as an evolving entity rather than a fixed stage. MST's localized dimensional "lumps" could, in principle, yield a continuum limit that mirrors LQG's discrete structure in certain low-dimensional regimes or embed spin networks in stable regions while letting extra dimensions emerge around matter. By unifying matter and spacetime from the outset, MST may also sidestep some of LQG's challenges regarding matter coupling and semiclassical limits. Moreover, MST remains compatible with LQG: future research might reveal how mixed dimensionality and division algebras in MST can help LQG recover general relativity more naturally or facilitate emergent particle physics. Conversely, MST can leverage LQG's insights into discrete structures and deltas to handle discontinuities that arise from mixed-dimensional geometries.

MST and Twistor Theory

Twistor Theory recasts spacetime points as algebraic entities in complex projective "twistor space", aiming to simplify both gravitational and gauge interactions via conformal symmetry and analytic methods. MST, with its mixed-dimensional approach, does not rely on twistors as the primary building block but similarly seeks a geometric unification where locality and dimensional structure arise naturally around matter. Although the two frameworks differ in mathematical language—twistors focus on complex-analytic geometry, MST on continuous manifolds with locally varying dimensions—they share an emphasis on reducing dependence on a fixed 4D background. MST could, in principle, be reinterpreted in twistor language: especially if localized, higher-dimensional "bubbles" turn out to correspond to specific configurations in twistor space. Hence, one can imagine a hybrid approach where twistor-based methods for scattering amplitudes or black hole horizons might mesh well with MST's dynamic geometry, offering new ways to compute or conceptualize local dimensional changes in strongly curved or highly energetic regions.

MST and String Theory

String Theory unifies gravity with quantum field theory by positing tiny vibrating strings in extra compactified dimensions. MST, on the other hand, treats higher dimensions not as universally compactified but dynamically created by localized energy lumps. While String Theory rests on supersymmetry (still unobserved) to balance bosonic and fermionic states, MST is agnostic on that front: it merely states that local patches of the MST fabric can assume higher dimensions when matter is sufficiently concentrated. One could envisage a scenario where MST's emergent dimensions serve as a mechanism for realizing, or partially emulating, the compact extra dimensions integral to String Theory, without requiring them to be globally present everywhere. Furthermore, MST's local dimensional variation might help explain the apparent absence of large-scale supersymmetry signals by restricting large extra dimensions to the most energetic regions. In this sense, MST's dynamic geometry could be compatible with string-based unification, offering a novel approach to the extra-dimension puzzle and potentially alleviating certain tuning or stabilization problems in String Theory's compactification landscape.

MST and Oppenheim's Post-Quantum Theory of Classical Gravity

Oppenheim's post-quantum theory explores the possibility that matter can remain fully quantum while gravity is treated as a classical field. A fundamental issue is ensuring consistency at the quantum–classical boundary—namely how quantum coherence or entanglement interacts with a classical spacetime metric. MST provides a perspective in which matter and geometry are not strictly separate in the first place: whenever matter lumps form, local geometry adapts by expanding the dimensionality of the MST fabric. In principle, this adaptability might offer a middle ground, where aspects of gravity remain effectively classical at large scales yet have an underlying quantum-geometry root in high-dimensional lumps. MST thus might serve as a unifying framework that accommodates a semi-classical viewpoint while circumventing the usual inconsistencies, since the geometry can locally "quantize" via extra dimensions where needed—effectively bridging quantum matter and classical gravitational fields.

Concluding Note on Quantum Gravity

While MST diverges from each of these established approaches in methodology and emphasis, mixed-dimensionality is not inherently incompatible with LQG, String Theory, post-quantum classical gravity, or Twistor Theory. Instead, MST could be viewed as a meta-framework capable of hosting or blending features from these theories: from LQG's discrete quantum geometry, to String Theory's extra dimensions, to post-quantum classical gravity's semi-classical coupling, to Twistor Theory's complex-analytic reparameterizations. The unifying notion is that matter-induced dimensional "lumps" may reconcile or at least offer new perspectives on challenges—such as how to incorporate matter fields, explain extra dimensions, handle quantum–classical boundaries, or capture geometric scattering processes—thereby suggesting fruitful directions for future cross-pollination among these diverse quantum gravity endeavors. Till

5.5 Normed Division Algebras and Quantum Systems

Each quantum system in MST theory has its own dimensions of freedom consistent with normed division algebras as itself and its environment are the result of large scale division of MatterSpaceTime. This insight explains the use of complex numbers in quantum mechanics, the reduction of Maxwell's equations using quaternions, and ongoing research into the Standard Model using octonions.

Complex Numbers in Quantum Mechanics

The use of complex numbers in quantum mechanics arises naturally in MST theory, where each quantum system follows the structure of normed division algebras. Complex numbers are the simplest example of a normed division algebra, corresponding to two-dimensional quantum systems.

- **Complex Numbers and the Schrödinger Equation:** The Schrödinger equation, which governs the evolution of quantum systems, is inherently complex-valued. This is consistent with the fact that quantum systems in MST theory follow the algebraic structure of complex numbers:

$$i\hbar \frac{\partial \psi(t)}{\partial t} = H\psi(t)$$

  where $\psi(t)$ is a complex-valued wavefunction.

Quaternions and Maxwell's Equations

Using quaternions, Maxwell's equations in electromagnetism can be reduced to a single equation, reflecting the underlying quaternionic structure of the fields.

- **Quaternion Representation of Maxwell's Equations:** In MST theory, electromagnetic fields can be represented as quaternionic-valued functions, leading to a compact formulation of Maxwell's equations. For example, the quaternionic field F can encode both the electric and magnetic fields:

$$\nabla F = J$$

  where J represents the source current. This quaternionic formulation is consistent with the higher-dimensional structure of quantum systems in MST theory.

Octonions and the Standard Model

Ongoing research in MST theory explores the possibility that the Standard Model of particle physics can be understood using octonions, the next level of normed division algebras after quaternions.

- **Octonionic Structure of the Standard Model:** The octonions, which have eight dimensions, may provide a natural framework for describing the complex interactions between particles in the Standard Model. The symmetry properties of octonions could explain the observed symmetries of the strong, weak, and electromagnetic forces.

- **Equations and Predictions:** The algebraic structure of the Standard Model could be encoded in octonionic fields, leading to predictions about particle masses, interactions, and possible extensions of the Standard Model. These predictions could be tested through high-energy experiments, such as those conducted at particle accelerators like the Large Hadron Collider (LHC).

5.6 Conclusion

MST theory offers a comprehensive framework that integrates quantum mechanics with higher-dimensional structures, resolving classical quantum paradoxes while providing a theory of quantum gravity. By extending dimensionality and leveraging normed division algebras, MST advances our understanding of the universe's quantum and classical behaviors.

6 Cosmology in MatterSpaceTime (MST) Theory

Cosmology in MatterSpaceTime (MST) theory presents a novel perspective on the evolution of the universe by integrating ideas from general relativity, quantum mechanics, Big Bang cosmology, and high-dimensional algebraic structures. This section elaborates on the key ideas of MST cosmology, focusing on the nature of time, the expansion & contraction of the universe, and the role of high-dimensional structures in cosmic evolution. This is depicted in the figure 3.

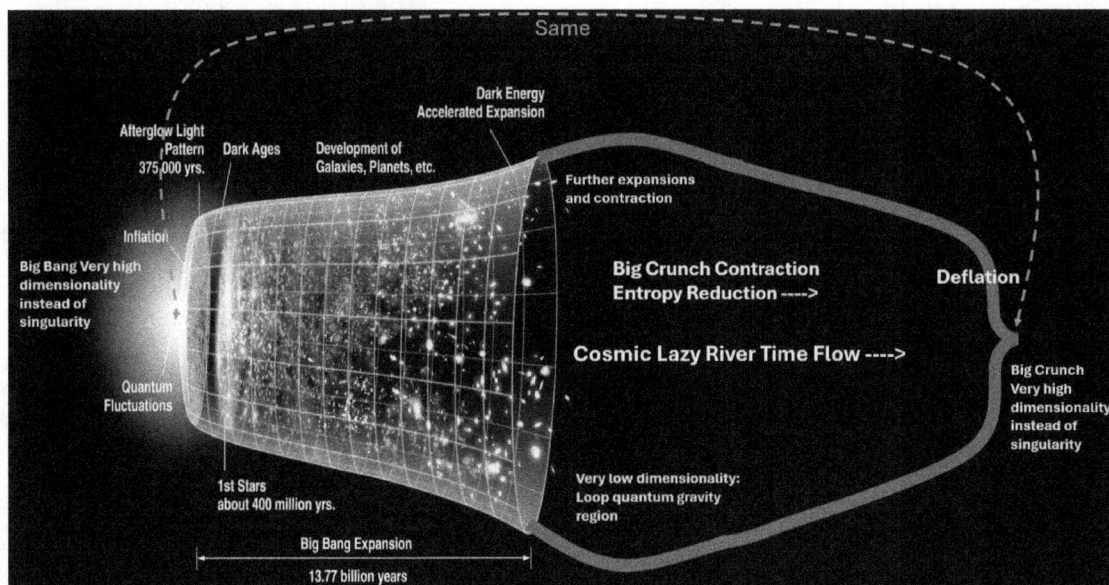

Figure 3: Cosmology in MatterSpaceTime (MST) Theory

6.1 The Block Theory of Time and the Arrow of Time

MST theory utilizes block theory of time which is supported by both general relativity and Big Bang cosmology. The block universe model views time as a dimension similar to space, where past, present, and future co-exist as a static "block."

Block Universe and Spacetime River Analogy

The universe can be imagined as a cosmic lazy river flowing through spacetime, with the Big Bang as its origin. In this analogy, the flow of the river represents the passage of time.

- **Arrow of Time**: The arrow of time can be understood as classical objects "going with the flow" of this cosmological river. Till contraction starts, the forward flow of time corresponds to the expansion of the universe, with entropy increasing over time. Once the contraction starts, forward flow of time corresponds to the contraction of the universe, with entropy decreasing over time.

6.2 Expansion of the Universe

As the universe expands, the cosmological river analogy implies that the spacetime fabric is stretching, thinning out as time progresses. This expansion is consistent with observations of the accelerating universe, driven by dark energy.

Thinning Out of MatterSpaceTime Fabric

In MST theory, the expansion of the universe is not merely an increase in the distance between galaxies but a "thinning out" of the MatterSpaceTime fabric itself. The dimensionality of spacetime may decrease as the universe expands, leading to a minimal-dimensional state where spacetime becomes increasingly sparse.

- **Minimal Dimensionality**: The concept of minimal dimensionality refers to a state where the universe reaches a point of maximum expansion, beyond which further expansion would lead to a collapse in the structure of spacetime.

- **Predictions**: This phase of the universe's evolution could be tested through precise measurements of the cosmic microwave background (CMB) radiation, gravitational waves, or the large-scale structure of the universe. Deviations from standard cosmological models at late times could indicate the onset of minimal dimensionality.

6.3 The Big Crunch and Entropy Reduction

According to MST theory, once the universe reaches a state of minimal dimensionality, it begins to contract, ultimately leading to a Big Crunch. This contraction phase corresponds to a decrease in entropy. In MST, the "size" of the universe is tied to how many distinct objects (or partitions) the MatterSpaceTime fabric has divided itself into. As the universe contracts, the number of these discrete objects diminishes, and with fewer overall configurations available, entropy naturally decreases. While the second law of thermodynamics dictates that entropy tends to increase, MST theory suggests that a cosmological phase transition (from expansion to contraction) will lead to opposite of second law of thermodynamics, making sign of entropy change dictated by second law of thermodynamics dependent on expansion or contractions of the Universe.

Big Crunch as Entropy Reduction

The contraction of the universe would lead to a concentration of matter and energy, reversing the entropy increase that occurred during the expansion phase. This reduction in entropy will lead to minimal entropy at Big Crunch. This reduction in entropy could be understood as a "reset" of the universe, leading to a new cycle of expansion and contraction. The cyclic universe is also predicted by Loop Quantum Gravity(LQG) and Conformal cyclic cosmology by Penrose.

- **Testing the Big Crunch Hypothesis**: Observations of large-scale cosmic structures and gravitational wave signals from collapsing regions could provide evidence for the onset of a contracting phase. Future experiments, such as those involving next-generation space telescopes and gravitational wave detectors, could offer insights into the dynamics of the universe during this phase. But it might be billions of years before this phase starts but we have to see if contraction happens in a smaller area in our lifetime.

6.4 The Cayley-Dickson Construction and the Big Bang

In MST theory, the Big Bang can be understood as the result of a super black hole leading to a super white hole that is packed via the Cayley-Dickson construction into a very high-dimensional structure. Loop Quantum Gravity also predicts white holes.

High-Dimensional Packing of early Universe and Validation

The Cayley-Dickson construction is a recursive process that generates higher-dimensional algebras, such as the quaternions and octonions, from lower-dimensional ones. In MST theory, this construction is applied to the universe's singularity, resulting in the packing of spacetime into a dense, very high dimensionality hyper-complex structure.

- **Cayley-Dickson Construction and High-Dimensionality**: The Cayley-Dickson construction allows for the creation of spaces with increasingly complex dimensions. The super black hole and white hole pair could be described as existing in an octonionic or even higher-dimensional space, where the packing of dimensions avoids the singularity with minimal entropy.

- **Inflation as Unpacking of Dimensions**: When this high-dimensional structure is "unpacked," it leads to the inflationary epoch of the Big Bang. The rapid expansion of the universe during inflation can be understood as the rapid unfolding of these high dimensions, which compactified the spacial slice of Universe during the high-dimensional phase.

- **Equations, Predictions and Validation**: The inflationary expansion could be described by a generalized form of the Friedmann equations that includes contributions from higher-dimensional terms:

$$\left(\frac{\dot{a}}{a}\right)^2 = \frac{8\pi G}{3}\left(\rho + \rho_{\text{higher-dim}}\right) - \frac{k}{a^2} + \frac{\Lambda}{3}$$

where $\rho_{\text{higher-dim}}$ represents the energy density associated with the unpacking of higher dimensions. Experimental predictions for MST include potential deviations from standard inflationary models, which can be probed through observations of the cosmic microwave background (CMB) and the universe's large-scale structure. One key signature might arise if the Cayley-Dickson "unwinding" process (during which the effective dimensionality of spacetime halves) imprints subtle anomalies in the CMB data. Although the CMB was emitted after inflation ended, these earlier dimensional transitions may have left traces in the form of unusual correlations or patterns. Among all MST predictions, this avenue—searching the already-collected CMB data for such anomalies—appears the most readily testable, making it a promising route for near-term empirical validation of MST. Above equation will be revisited, and specific predictions of CMB anomaly will be documented in a separate paper in the future.

6.5   Observable Universe and Matter-Spacetime Separation

As the high-dimensional structure of the universe unpacks during inflation and expands further, matter particles appear to separate from the spacetime fabric after inflation, leading to the observable universe.

Matter-Spacetime Separation

In MST theory, matter and spacetime are not fundamentally separate entities but arise from the same underlying structure. After the inflationary epoch, further expansion of the universe causes matter particles to appear as though they are distinct from the spacetime fabric, leading to the emergence of classical objects and fields.

- **Matter as a Manifestation of Spacetime**: The separation of matter from spacetime in dimensionality can be described as a phase transition, where the underlying unified structure of MatterSpaceTime differentiates into observable matter and the surrounding spacetime. This phase transition could be modeled using field equations that couple matter fields to the spacetime metric.

- **Testing Matter-Spacetime Separation**: Experimental tests of this idea could involve searching for signatures of this phase transition in the early universe, such as deviations from standard models of particle physics or cosmology. High-energy particle accelerators and cosmological observations could provide clues about the nature of this separation.

6.6 Conclusion and Experimental Predictions

Cosmology in MST theory provides a novel framework for understanding the evolution of the universe, from the Big Bang to the present day, and beyond. By integrating ideas from general relativity, quantum mechanics, and high-dimensional algebra, MST theory offers a unified perspective on the nature of time, the expansion and contraction of the universe, and the role of high-dimensional structures in cosmic evolution.

By further exploration and making specific predictions, MST theory has the potential to offer new insights into the fundamental nature of the cosmos and the interplay between matter, spacetime, and higher dimensions.

7 Philosophy: Materialism vs. Idealism in MatterSpaceTime Theory

The philosophical debate between materialism and idealism has persisted for millennia in Western philosophy, dating back to the earliest philosophical inquiries in ancient Greece. Materialism posits that matter is the fundamental substance of reality, while idealism asserts that ideas or consciousness are the primary reality. This debate has shaped much of Western philosophy and has significant implications for understanding the nature of existence, consciousness, and reality itself. MatterSpaceTime (MST) theory offers a novel resolution to this debate by positing that both matter and ideas emerge from a single, more fundamental entity—MatterSpaceTime itself. This section elaborates on how MST theory reconciles materialism and idealism through the concept of "idea emergence," critically comparing this view to existing philosophical positions, and discussing the importance and impact of resolving this debate.

7.1 Historical Context of the Materialism vs. Idealism Debate

The materialism vs. idealism in Western philosophy debate can be traced back to the pre-Socratic philosophers of ancient Greece. Materialists like Democritus argued that everything in the universe is composed of indivisible atoms, while idealists such as Plato emphasized the primacy of abstract forms and ideas. This debate continued through the ages, with materialism taking center stage during the Enlightenment and the rise of modern science, while idealism found its strongest proponents in philosophers like Berkeley, Hegel, and Kant.

The debate remains relevant today, particularly in the context of modern physics, neuroscience, and artificial intelligence(AI), where questions about the nature of consciousness and the material world are central. Resolving this debate is crucial because it influences our understanding of reality, the nature of mind and matter, and the relationship between the physical and mental aspects of existence.

7.2 Importance of Resolving the Debate

Resolving the materialism vs. idealism debate is important for several reasons:

- **Philosophical Clarity**: The debate touches on fundamental questions about the nature of reality and existence. By resolving it, we can achieve greater clarity about whether reality is primarily material, mental, or something else entirely.

- **Impact on Science and Technology**: The debate also has implications for scientific inquiry. Understanding whether consciousness is a material phenomenon or something that emerges from non-material components could influence research in fields like artificial intelligence(AI), quantum mechanics, and neuroscience.

- **Ethical and Social Implications**: The resolution of this debate could also impact ethical and social perspectives. For example, materialist views often lead to a more deterministic outlook on life, whereas idealist views may emphasize the importance of mental and spiritual aspects of existence. Reconciling these views could lead to a more balanced approach to issues such as free will, personal identity, and morality.

7.3 Idea emergence in MatterSpaceTime

In MST theory, MatterSpaceTime possesses the intrinsic property of "Idea emergence." This means that various forms, whether geometrical (e.g., shapes, structures) or process-like (e.g., events, interactions), naturally emerge from the fundamental substrate of MatterSpaceTime. The process of idea emergence involves the division of the single entity, MatterSpaceTime, into distinct components, which then manifest as different forms of reality. These components can be spatially separated parts, wave-like structures, or more abstract elements, such as the terms of a Fourier series.

- **Geometrical and Spatial Forms**: Geometrical forms, such as shapes, structures, and spatial configurations, emerge as MatterSpaceTime divides itself into spatially separated parts. These forms represent the observable geometries in the physical world, arising from the fundamental divisions within MatterSpaceTime.

- **Wave-like and Continuous Forms**: Wave-like forms, where components are continuous, oscillatory structures, also emerge from the division of MatterSpaceTime. These components, analogous to waveforms in physics or Fourier series in mathematics, represent continuous patterns that arise from the division of the single entity into complex, interacting waves.

- **Abstract and Non-Obvious Components**: MST theory also accommodates the emergence of more abstract forms, such as the components of a Fourier series. These components are not immediately obvious in the physical sense but represent the underlying structures that combine to produce observable phenomena.

7.4 Emergence of Ideas and Consciousness

Ideas, including complex forms such as consciousness, are also seen as emergent phenomena in MST theory. Just as matter arises from the division of MatterSpaceTime into spatial and wave-like components, so too do ideas and consciousness emerge from this single entity through the division of its less obvious, more abstract components. This view challenges the traditional dichotomy between materialism (which holds that matter is fundamental) and idealism (which holds that consciousness or ideas are fundamental) by suggesting that both emerge from the same underlying substrate—MatterSpaceTime.

Consciousness as Emergent Form

In MST theory, consciousness is not a separate entity from matter but rather an emergent form that arises from the complex divisions within MatterSpaceTime. These interactions may involve both obvious components, such as neural activity in the brain, and more abstract components, such as those represented in the frequency domain (e.g., neural oscillations modeled through Fourier analysis).

- **Emergence of Complex Forms**: Consciousness, as a complex form, emerges from the interaction of both spatially separated parts and wave-like or abstract components within MatterSpaceTime. The higher the complexity of the system (e.g., a brain or neural network), the more sophisticated the emergent form (e.g., consciousness).

7.5   Resolving the Materialism vs. Idealism Debate

MST theory provides a resolution to the traditional philosophical debate between materialism and idealism by positing that both matter and ideas (including consciousness) emerge from a single fundamental entity: MatterSpaceTime. This view suggests that neither materialism nor idealism is more fundamental than the other, as both are simply different manifestations of the same underlying reality.

Materialism and Idealism as Emergent Dualities

In MST theory, materialism and idealism are seen as emergent dualities rather than competing worldviews. Both are valid descriptions of reality, but they describe different aspects of the same fundamental substance. Matter represents the spatial and wave-like aspects of MatterSpaceTime, while ideas and consciousness represent the more abstract and processual aspects that emerge from the complex interactions of components.

- **Materialism as Spatial and Wave-like Emergence**: From the perspective of materialism, the physical world, including matter and energy, emerges from the spatial divisions and wave-like configurations of MatterSpaceTime. The physical properties of objects and fields are the result of the way MatterSpaceTime divides itself into spatial parts and waveforms.

- **Idealism as Abstract Emergence**: From the perspective of idealism, ideas, consciousness, and mental phenomena emerge from the more abstract components of MatterSpaceTime, such as those represented in Fourier series or other mathematical structures. These non-physical properties arise from the dynamic interactions and relational patterns of components that are not always directly observable.

Free Will

As discussed in the quantum gravity subsection before while discussing gravity-induced decoherence, the emerged form has an inertia to maintain its form/pattern and this in turn influences the behavior of underlying layers. Free will is similar in the sense a conscious decision made by the mind can influence and control the underlying the body to implement the decision. It could be decision making is emerging out of the brain but once the decision is made it can be carried out by the body. Unlike classical determinism, in MST we have influence/control going up and down the levels of reality. Hence MST theory has unique perspective on top down controls in addition to bottom up emergence justifying the existence of free will.

Eventhough classical physics is fully deterministic, since quantum objects are the underlying components, non-deterministic quantum nature accommodates possibilities. Quantum probability of the consciously chosen option's quantum wavefunction can be increased by the conscious individual by setting up the environment appropriately. In MST classical determinism gives way to quantum possiblities by increasing the dimensionality of the underlying quantum objects appropriately which in turn provides independent variables which are controllable by the individual. This is how Free Will is defended within MST.

Qualitative and Quantitative Unity

MST theory suggests that materialism and idealism both emerge qualitatively and quantitatively from the same fundamental entity. MatterSpaceTime is a quantitatively single entity that divides itself into components, from which both matter and ideas emerge. Qualitatively, matter and ideas are different expressions of MatterSpaceTime's potential to generate forms. Quantitatively, both matter and ideas are governed by the same underlying principles of division and interaction within MatterSpaceTime, whether these components are spatial parts, waves, or abstract mathematical entities.

- **Qualitative Unity**: The qualitative unity of materialism and idealism lies in their shared origin in MatterSpaceTime. Whether we are observing physical matter, wavelike phenomena, or abstract ideas, all are emergent properties of the same fundamental substance, differentiated only by the way they manifest in space, time, and mathematical structure.

- **Quantitative Unity**: The quantitative unity of materialism and idealism can be understood through the framework of MST theory, where MatterSpaceTime is a single entity that quantitatively divides itself into different forms. The rules governing this division apply equally to the emergence of both material and ideal forms. This can be represented using algebraic structures and wave decomposition methods, such as Fourier analysis, that describe the division of MatterSpaceTime into various components that manifest as different forms of reality.

7.6 Critical Comparison with Existing Philosophies

MST theory's approach to resolving the materialism vs. idealism debate can be critically analyzed and compared to various existing philosophical viewpoints.

Comparison with Classical Materialism

Classical materialism posits that matter is the fundamental substance of reality, and all phenomena, including consciousness, are reducible to material interactions. MST theory differs from classical materialism by proposing that matter itself is an emergent property of a more fundamental substrate—MatterSpaceTime. Moreover, MST theory expands on materialism by considering wave-like and abstract components as integral to the emergence of material forms.

Comparison with Classical Idealism

Classical idealism posits that consciousness or ideas are the fundamental substance of reality, and the physical world is a manifestation of the mind. MST theory challenges this view by proposing that ideas and consciousness are emergent properties of MatterSpaceTime, just as matter is. This suggests that consciousness arises from both obvious and non-obvious components, such as waveforms and abstract mathematical structures, making it a complex emergent form rather than a fundamental substance.

Comparison with Dualism

Dualism holds that mind and matter are two distinct substances that interact with each other. MST theory, by contrast, posits that mind (ideas) and matter are not distinct substances but rather emergent forms of the same fundamental entity. This monistic view avoids the problems of dualism by unifying the two under a single framework, where both emerge from the same substrate but in different forms, including spatial parts, waves, and abstract components.

Comparison with Neutral Monism

Neutral monism [12] is a philosophical stance—prominently advocated by William James, Ernst Mach, and Bertrand Russell—holding that the fundamental "stuff" of reality is neither purely mental nor purely physical, but rather a neutral substrate that can manifest as either mind or matter depending on context or perspective. MST resonates with this view by positing a single entity (MatterSpaceTime) that underlies all phenomena. Instead of treating matter and consciousness as separate substances (dualism) or reducing one to the other (materialism vs. idealism), MST treats dimensional variability and energy distributions within a unified fabric as the source of both physical (spatiotemporal) and mental or informational qualities. In this sense, MST can be viewed as "neutral" in that it does not privilege the material or the mental but instead provides a structural foundation—mixed-dimensional geometry—that can give rise to both. Such an approach aligns well with the core motivation of neutral monism: to explain the apparent dualities of experience and physicality by appealing to a more fundamental, all-encompassing substrate.

Comparison with Emergentism

Emergentism is a philosophical perspective that holds that higher-level phenomena (e.g., consciousness) emerge from lower-level physical processes but are not reducible to them. MST theory aligns with emergentism by suggesting that both matter and ideas emerge from MatterSpaceTime, but it extends this view by proposing that both are the result of the division of a single, quantitatively unified entity, rather than treating them as separate domains. In addition, existence of free will is defended in MST theory using the mechanism of forms in higher emergent layers influencing lower layers all the way down to quantum layer.

7.7 Conclusion: A Unified View of Reality

MatterSpaceTime theory offers a unified view of reality, where both materialism and idealism are seen as emergent aspects of the same fundamental entity. By framing both matter and ideas as forms that emerge from the division and interaction within MatterSpaceTime, MST theory provides a novel resolution to the long-standing debate between materialism and idealism. This perspective opens new avenues for exploring the nature of reality, consciousness, and the interplay between the physical and mental aspects of existence.

8 Conclusion and Future Directions

8.1 Conclusion

MatterSpaceTime (MST) theory offers a bold and innovative framework that redefines our understanding of the relationship between matter, spacetime, and dimensionality. By proposing that matter and spacetime are unified into a single mixed-dimensional fabric, MST theory challenges the traditional view of matter as something that exists within spacetime. Instead, it suggests that matter is a manifestation of localized energy concentrations that create dimensional variations within spacetime itself. This approach opens new avenues for addressing some of the most pressing problems in theoretical physics, including the unification of quantum mechanics and general relativity. Table 1 compares traditional and MST methods.

One of the most profound implications of MST theory is its treatment of extra dimensions. Unlike string theory, which requires compactification of extra dimensions, MST theory posits that dimensions are created dynamically by the presence of matter and energy. This eliminates the need for compactification and provides a more intuitive understanding of how dimensions and matter co-exist.

MST is compatible with quantum gravity theories like Loop Quantum Gravity which describe spacetime in discrete in most fundamental level. Since MST does not require discreteness it can also work with theories such as "Postquantum theory of classical gravity" described in [11].

Challenge	Traditional Methods	MST (MatterSpaceTime)
Nature of Spacetime (Fundamental vs. Emergent)	Debated: General Relativity (fundamental), String and Loop Quantum Gravity (LQG) (Emergent)	Emergent: Mixed dimensional fabric which provides the emergent dynamic dimensionality, topology, local symmetries and distinguishes itself from emptiness background which has no properties or symmetries; Metric emerging from mixed dimensional discrete entities as in LQG or from string vibrations
Quantum Gravity	String Theory and LQG etc.	MST integrates spacetime with matter, creating a unified framework extends/generalizes other quantum gravity theories such as LQG.
Singularities (Black Holes, Big Bang)	Predicted by General Relativity, avoided in LQG	Avoided through higher-dimensional transitions, dynamic adaptation, and quantum effects similar to LQG
Nature of Time	Linear (classical absolute time), Non-linear (block universe as in special/general relativity)	Non-linear: Block Universe
Dimensions of Spacetime	Fixed (3 space + 1 time in relativity), Extra dimensions in string theory	Variable: Dimensionality can vary across spacetime fabric, influenced by energy/matter
Causality and Correlation	Causal structure in General Relativity, quantum correlations (entanglement)	Both causal and correlative properties emerge, with quantum systems influencing local spacetime structure; both bottom up/top down causations/correlations
Nature of Matter and Energy	Separate but convertible entities influencing spacetime curvature	Integrated with spacetime, matter and energy interacting with the spacetime fabric influencing dimensionality/curvature/metric/topology
Continuity vs. Discreteness	Continuous spacetime (General Relativity), Discrete spacetime (Loop Quantum Gravity)	Mixed: Continuous fabric with mixed dimensionality with discontinuities when dimensionality changes.
Unification of Forces	Attempts through String Theory, Quantum Field Theory	Integrated framework where matter and spacetime are unified, potentially simplifying the unification of forces

Table 1: Comparison of Traditional Methods and MST in Addressing Challenges in Spacetime

The mathematical framework underlying MST theory, including mixed-dimensional manifolds, fiber bundles, and algebraic topology, provides a solid foundation for exploring these new ideas. Additionally, MST theory has far-reaching applications in areas such as quantum gravity, cosmology, and even AI/ML(machine learning). By incorporating mixed-dimensionality into these fields, MST theory has the potential to offer new insights into the behavior of quantum fields, the formation of black holes, and the efficiency of optimization algorithms.

8.2   Philosophical Implications

The philosophical implications of MST theory are equally profound, as it challenges the traditional dualism between matter and space. In this theory, matter is not a separate entity that exists within spacetime; instead, matter and spacetime are fundamentally intertwined. This perspective aligns with a form of ontological monism, where the universe is composed of a single substance—MatterSpaceTime—rather than distinct entities of matter and space.

This unification of matter and spacetime also blurs the boundaries between the physical and metaphysical, suggesting that what we perceive as distinct objects or events are merely manifestations of varying dimensionalities within a continuous fabric. In this way, MST theory offers a potential bridge between materialism and idealism, proposing that the reality we experience is a product of dimensional variations rather than independent objects existing in a fixed space.

MST theory's emphasis on the dynamic and emergent nature of dimensions also resonates with process philosophy, which views reality as constantly becoming rather than a fixed state of being. This perspective encourages a rethinking of the foundational assumptions in both physics and metaphysics, promoting a worldview where change, emergence, and interconnection are central.

8.3   Future Directions

While MST theory presents a promising new framework, much work remains to be done. Future research should focus on developing the mathematical tools needed to fully explore the implications of mixed-dimensionality. This includes further extensions of tensor calculus, integration theory, and algebraic topology to handle varying dimensions across regions of spacetime.

Additionally, MST theory offers several avenues for experimental validation. For example, the theory's predictions about dimensional variations in high-energy environments, such as near black holes or during the early universe, could be tested through astrophysical observations. Future papers will fine-tune these predictions and document ways to validate the predictions.

Finally, MST theory's applications in AI/ML(machine learning) offer exciting possibilities for improving algorithmic efficiency and performance. Future research will explore how mixed-dimensionality can be leveraged to optimize training of neural networks and other machine learning models.

In conclusion, MST theory is a versatile and comprehensive framework that holds great promise for advancing our understanding of the universe. By unifying matter, spacetime, and dimensionality, MST theory provides a new lens through which to view the fundamental nature of reality, offering insights that could bridge the gap between quantum mechanics, general relativity, and beyond. Its philosophical implications, challenging long-held distinctions between matter, space, and time, invite further exploration of the metaphysical foundations of physics, potentially reshaping our understanding of reality itself.

References

[1] Minkowski, H., *Space and Time, Published in: Jahresbericht der Deutschen Mathematiker-Vereinigung, 1908.*

[2] Einstein, A., *The Foundation of the General Theory of Relativity, Annalen der Physik, 1916.*

[3] Hawking, S., & Ellis, G. F. R., *The Large Scale Structure of Space-Time, Cambridge University Press, 1973.*

[4] Penrose, R., *The Road to Reality: A Complete Guide to the Laws of the Universe*, Vintage Books, 2004.

[5] Penrose, R., *On the Gravitization of Quantum Mechanics 1: Quantum State Reduction, Foundations of Physics, 44(5), 557–575, 2014.*

[6] Greene, B., *The Elegant Universe: Superstrings, Hidden Dimensions, and the Quest for the Ultimate Theory, W. W. Norton & Company, 1999.*

[7] Dieks, D., & Redei, M. (Eds.), *The Ontology of Spacetime, Philosophy and Foundations of Physics Series (Vol. 1), Elsevier, 2006.*

[8] Bianchi, E., & Martin-Dussaud, P., *Causal Structure in Spin-Foams, Loop Quantum Gravity: A Themed Issue in Honor of Prof. Abhay Ashtekar, 2023.*

[9] Petkov, V., *From Illusions to Reality: Time, Spacetime, and the Nature of the Universe, Springer, 2023.*

[10] Erbin, H., *String Field Theory, Lecture Notes in Physics (Vol. 980), Springer, 2021. ISBN: 978-3-030-65320-0.*

[11] Oppenheim, J., *A Postquantum Theory of Classical Gravity, Physical Review X, 13(4), 041040, 2023.*

[12] Stubenberg, L., & Wishon, D., *Neutral Monism, The Stanford Encyclopedia of Philosophy*, E. N. Zalta & U. Nodelman (Eds.), Spring 2023 Edition, *Metaphysics Research Lab, Stanford University*, 2023. *URL: https://plato.stanford.edu/archives/spr2023/entries/neutral-monism/.*

8   Human as a Time Structure

Nikola Pirovski

Abstract   The human body, in the light of Minkowski's theory, can be interpreted as a four-dimensional object that exists not only in space but also in time. This reminds us that our existence is not simply momentary, but is a full-fledged part of the complex structure of spacetime. These ideas expand our perspectives on how we understand the human body – not simply as physical matter, but as a form inextricably linked to the time and space in which we exist. Humans, as temporal structures, are influenced by a complex interplay of physical laws, biological rhythms, psychological perceptions, cognitive processes, and socio-cultural factors. Understanding these dimensions provides insight into how we experience, measure, and interact with time, but also into its nature and the place for the free will. The more conscious we become about the scientific order of the world, the less identifications seem to us as free to choose. However the dimensions of the human existence leave us with at least one in which we are free from determination and able to make our own subjective mind.

Keywords: human, space time, STHB, free will

1   Introduction

Anthropology explores all aspects of the human existence. The physics discoveries and theories are fundamental for the understanding of the human existence and morphology (Capra, 1999). On the other hand, the human morphology could be a source of information both for the basic underling physical laws and the most complicated structures. Exploring the human being as a time structure involves examining how biological, psychological, and cognitive processes interact with and are influenced by time. Examples of Biological Time Structures are: Circadian Rhythms, Cellular Processes, Hormonal Cycles ect. Psychological Time Perception involves Subjective Time, Temporal Integration and Cognition. Memory, decision-making and planning are intrinsically linked to time (Grondin, 2010). There is also a Sociocultural Constructs of Time. Different cultures perceive and value time differently which shape how individuals experience and management of time (Hall, 1973). Some cultures have a linear view of time, emphasizing schedules and punctuality, while others may have a more cyclical, event-oriented or person-oriented perception of time (Hofstede, 1980; Levine, 1997).

The physics discoveries and theories are fundamental for the understanding of the human existence and morphology. New physics theories emerged during the last century and radically changed the scientific understanding of the nature. They still need to be incorporated in the understanding of human morphology. My professor in neuromorphology, my professor in cytology both have really difficult time with the full contemplation of two of the biggest scientific efforts of our time in understand the human – Connectome project[1]

---

[1] The Connectome Project aims to map the intricate network of neural connections in the human brain. By

Kyley Ewing (Ed.), *Spacetime Conference 2024. Selected peer-reviewed papers presented at the Seventh International Conference on the Nature and Ontology of Spacetime, 16 - 19 September 2024, Albena, Bulgaria* (Minkowski Institute Press, Montreal 2025). ISBN 978-1-998902-44-6 (softcover), ISBN 978-1-998902-45-3 (ebook).

and the Human cell atlas project.[2] They had shared their unsatisfaction, because they got lost in the varieties and could not find the pattern. Important theories that are still not integrated in the understanding of human body are hologram, fractal, quantum, synergy etc. It is time for the human anatomy science to make a paradigm shift from independent, dead and immobile body, to alive and moving organism which correlates with the physics laws.

Human anatomy science has not changed much during last century in contrast to mathematics and physics. Many phenomena remain unexplained and the description is still positioned it in three dimensional space. The studies are mainly of a dead body, which lacks the most important properties of the living body, and rejects to incorporate theoretical constructs that are already empirically proven. Spiral theory of the human body is aiming to challenge this paradigm with explanation of the living human body in multi-dimensional spacetime. While the new ideas in physics need geometrization, the ideas in human anatomy need their mathematical formalism.

Hermann Minkowski, a German mathematician and physicist, is known for his contribution to the understanding of time and space through the concept of four-dimensional spacetime. His work is fundamental to Einstein's theory of relativity. If we consider the human body in relation to Minkowski's physical theory, we can make interesting interpretations related to the way the body exists and interacts in time and space.

Exploring the concept of humans as a time structure within the framework of physics involves different directions that need to be merged together: 1. human existence, 2. perception of time. They could be interpreted through the principles of physical theories.

The main questions about the properties of time are:

1. Dimension and Framework:

- Fundamental force, or / emerging from more basic process.

- One dimension – a measure of the sequence and duration of events, or / Four dimensions that combines space and time into a unified entity.

2. Relationship and relativity:

- Separate from space in classical physics and Absolute (the same for everyone), flows uniformly for all, or / time and space are interrelated, relative and the flow of time can change depending on the observer's velocity and gravitational field (as described by relativity).

---

studying how neurons and their pathways are organized and interact, the project seeks to understand brain function, cognition, and behavior, as well as the neural basis of diseases like Alzheimer's and schizophrenia. The most notable initiative is the Human Connectome Project, which combines advanced imaging technologies and computational methods to create a comprehensive map of the brain's structural and functional connectivity.

[2]The Human Cell Atlas Project is an international initiative to create a detailed reference map of all the cells in the human body. By cataloging the diversity of cell types, their functions, and their interactions, the project provides a foundation for understanding human biology, health, and disease at the cellular level. This atlas aims to guide advancements in precision medicine and improve treatments for various conditions by offering insights into how cells behave in healthy and diseased states.

3. Purpose

- Physiologically necessary, or / just pathological temporal illusion.

While the answers seem incompatible in a single frame, the Scale theory could easily accommodate opposite functions at the same time, but on a different scale. There are many examples of properties that emerge or change due to change in scale. Time is also such an example, with the Minkowski experiments proving it is fundamentally integrated with space, and the psychology and psychiatry, that it is linear and emerging property. This way we could say that subjective time is a dimension allowing us to sequence events, measure durations, and understand the progression of change. It is the ongoing progression of existence and events that move from the past through the present into the future, experienced like in classical physics, considered separate from space, treated as a constant, universal entity that flows the same for all observers. However this illusionary perception is a construct enhanced over the reality of spacetime, which is a four-dimensional continuum that combines the three dimensions of space (length, width, height) with the one dimension of time into a single framework. Events are located not just in space, but also in time, can bend or curve in response to mass and energy (as described in general relativity), only present exists, and events are integrals. Practical medical observation proof, that time perception is mainly a result of normal cyclic biological process and paramount for the normal psychology (Wirth et. al., 2024).

2   Goal

Evaluating the human morphology as a source of information for the properties of spacetime. The technological advances allowed us to reach beyond many of the limitations we originally have as biology structures. Is this process going to end, already finished or limitless? When are we going to notice if it has ended and it is time to sort out what we know an choose not only what it is, but what we need it to be. If the science requires us to reduce the number of assumptions to minimum, could we limit it to only one?

3   Materials and methods

Available scientific information on human morphology and spacetime. Descriptive analysis and system thinking.

A system is a group of related elements that work together as a whole and is described by its: center, border, elements and hierarchy. If the border of the Universe (currently the Plank space and astronomical horizons) and too far away, and the elements are beyond count and calculation, we should focus on the center. Center could be enough to defying the whole system. The center in this publication is paced on the Human, which is available close to us, abundant in variety and susceptible for study.

4   Results

- Human scale

Humans, as time structures, are influenced by a complex interplay of physical laws, biological rhythms, psychological perceptions, cognitive processes, and sociocultural factors. Understanding these dimensions provides insight into how we experience, measure, and interact with time but also about its nature and the place for free will. Human activity is the most complex implementation of time and could give us a clue on its properties.

A human is natural part of the Universe and could be measured compared to the other structures in it. Orders of magnitude- length: Smallest- Subatomic- 10^{-36} to 10^{-33}; Middle: 1 (10^0); Larger: 10^{27}. Orders of magnitude- time: Smallest: Plank - 10^{-44}; Middle: 10^1; Larger: 10^{30} and onward. A human is somewhere in the middle of this scales of their making, commonly with life expectance from 1 s. to 100 years and body length from 100 micrometers to 2,75 meters. The middle position in the scale is beneficial for reaching to the lower border and the upper border, providing maximal opportunity for horizons. Three things remain beyond its capabilities of observation and explanation: 1 . the universe outer border; 2. the quantum world, as the universe smallest border; and 3. themselves. If we understand any of it, the other two would be easier to explain. While we rely on machines to explore the most of the universe, for the self-exploration the best tool is another human. The quantum and astrophysics research are resource and time demanding, but the human interactions are abandoned and fast. The technological advances allowed us to reach beyond many of the limitations we originally have as biology structures. Is this process going to end, already finished or limitless? My view is that it is at least in stagnation. If the science requires us to reduce the number of assumptions to minimum, could we limit it to only one?

If I could paraphrase the Hempel pseudo dilemma to include the human being, it would say: the thesis of physicalism is either false or empty if they do not account for the human in it.[3]

According to Minkowski, every body or object is an integral part of spacetime. Minkowski considers space and time not as separate entities, but as an inseparable four-dimensional structure (three spatial dimensions and one temporal). The human body can be interpreted as an object that has a form not only in space, but also in time. The body does not exist only at a given moment, but as a "path" or "trace" in spacetime – from birth to death. This means that our existence is a four-dimensional form stretched across time. In Minkowski's theory, the events that are accessible to a given body are limited by the light cone – the region in spacetime where information or motion can reach, according to the limit of the speed of light. That also poses some limits on the human perforce. The human body can be considered as limited in its actions by the laws of physics – no part of the body or a signal sent by it (e.g. a nerve impulse) can exceed these limits. The mass and energy of the body: According to the Minkowskian interpretation, all mass is also a form of energy. The human body is not only a material object, but also a concentration of energy that interacts

---

[3]It suggests that physicalism, as a thesis, must go beyond abstract reductionism and fully integrate the rich, multidimensional aspects of being human. Otherwise, it is either incorrect (if it ignores the human entirely) or unsatisfactory (if it trivializes humanity by reducing it to mere physical processes).

with spacetime through gravity and motion. Although small, the mass of the human body affects spacetime by distorting it (according to the general theory of relativity, based on Minkowskian space). The body clock (for example, the biological rhythm) is dependent on movement and the environment. If the human body moves at a very high speed (close to that of light), time for it will flow more slowly compared to an observer at rest (time dilation). The aging of the human body, if studied through Minkowski's theory, can be considered as a phenomenon related to its movement through spacetime. A person's life can be viewed as a unique line (trajectory) in four-dimensional spacetime, determined by their actions, movements and choices. The human body does not exist in isolation – it is connected to its environment through gravity, motion and energy flows. This theory resonates with the classical Theseus' paradox and supports the Heraclitus solution.

- Human time dimensions are presented as a table. Fig.1.

While describing the human body most common approach is to position it as a solid body in three dimensional space (not coherent with relativity theory). The psychology provides us with many different questionnaires to evaluate the mental condition which vary in the number of tested categories. However the basic psychiatry interview adds 11 areas of interest or we could consider them as psychological dimensions. Psychomotorics, consciousness, attention, instinct, perceptual-representative activity, will, thinking, memory, emotions, intellect and insight. There is also the social dimension. All dimensions could not be separated and should be interrelated if we need to have a human being. Consciousness emerges from the interaction between unconscious processes (automatic functions, like breathing or background thoughts) and the focused, deliberate processes of awareness.

Category	Scientific field	Time measurement
Human Center	Philosophy: Human center	Anthropic principle
Elementary Particles	Physics: Physics laws	Isotopes stability

Each of these factors can affect how we perceive the passage of time. Time can seem to pass more quickly or slowly depending on the context and activities we are engaged in (Liu, 2024). The immune, endocrine, and nervous systems interact dynamically to regulate the body's response to internal and external stimuli over time. These interactions are mediated by cytokines, hormones, and neurotransmitters, creating a communication network. Immune System releases cytokines that influence the nervous system by altering behavior (e.g., fatigue, fever) and modulating endocrine responses. Endocrine System secretes hormones (e.g., cortisol, adrenaline) in response to stress or circadian rhythms, which regulate immune activity and influence neural functions. Nervous System modulates immune responses through the autonomic nervous system (e.g., the vagus nerve) and directly impacts endocrine hormone secretion. These systems synchronize with circadian rhythms to maintain homeostasis. Disruptions in their temporal coordination (e.g., chronic stress or sleep deprivation) can lead to dysregulation, increasing susceptibility to diseases.

Category	Scientific field	Time measurement
Cell Structures	Molecular biology: Membranes, cytocenter, chromosome telomeres	Molecular stability
Tissue Structures	Histology: Local hormones, cell contacts	cells' life span, biological age
Systems: Nervous system	Neurobiology, neurology, psychology, psychiatry	Attention, memory, and emotional state.
		Reticular Formation: Controls awakening and sleep cycles
		Nucleus Tractus Solitarius: Programmed death area
		Limbic Formation: Emotional regulation and memory processing, connection to circadian rhythms
		Small Brain (Cerebellum): Automatic motor functions; Fast movements acts on motor circuits of the cerebellum for millisecond-level events
		Pineal Gland: Regulation of day/night, seasons and sexual development cycles
		Medial Prefrontal Cortex: Constant focus on past and future; imagination
		Parietal Cortex: Informational integration, timeless perception, controlled cognitively, formed by parietal and prefrontal areas linked to attention and memory, being responsible for periods of minutes.
Other systems	All other systems participate with their unique rhythm, studied by different medical specialties	Autonomic nervous system, hearth rhythm, enteral movements, osteoporosis ect.
Apparats and Organism	Anatomy, physiology and psychiatry: System hormones, electromagnetic signaling, sensory information, circadian rhythms, consciousness, altered states of consciousness.	Regulation of complex biological processes, integration of sensory and hormonal signals. Number of conscious choices.
Society	Anthropology	Calendar

Fig.1. Dimensions of human time perception and regulation.

5   Biology models

Biology processes are automated, mostly unconscious repeated cycles. The curvature of spacetime gives them a spiral structure. The spiral structure of biological processes is a discrete function. Normal biological cycles facilitate normal psychological experience.

Phycology could alter the biological cycles via the psychosomatic connection.

Example: The skeletal system is the slowest to replace "all star dust" that built it and this process takes around 7 years. Behavior could alter this process and keep the balance for a long time, or lead to misbalance in the loss and accumulation of bone tissue and early osteoporosis.

Biological age seldom matches the calendar age. Chronological age is the number of years you've been alive, while biological age refers to how old your cells and tissues are based on physiological evidence. A person with a lower biological age compared to their calendar age may have a lower risk of age-related diseases and a longer healthspan, whereas the reverse might indicate accelerated aging or heightened disease risk. This distinction is crucial in fields like personalized medicine, geriatrics and anti-aging research. In some societies, calendar age heavily influences social roles, retirement age, or perceived ability. Recognizing biological age may challenge these assumptions, showing that individuals of the same chronological age can have vastly different physical and cognitive capacities. This could align with the idea that time is not perceived equally by all observers, as suggested by relativity. The divergence between biological and calendar age in physics contexts, emphasizes that the human body is not merely a passive observer but a participant in spacetime phenomena. Biological age reflects the accumulation of entropy and damage within a system, while calendar age does not account for these internal processes. While the biological age is more close to the objectiveness of spacetime, the calendar age is a social construct that is more suitable for measurement on a different scale, like generations and nations. If biological age is influenced by environmental spacetime factors, it could offer insights into how life interacts with the curvature of spacetime, gravity, and other relativistic effects. The distinction supports philosophical explorations of time, such as those proposed by Bergson and Merleau-Ponty, who emphasize the difference between lived time and objective time. Biological age adds a concrete dimension to this philosophical distinction, grounded in physical and biological processes. This distinction underscores the necessity of integrating physics and biology to understand time and spacetime in a holistic way.

Quantum biology is the study of applications of quantum mechanics and theoretical chemistry to aspects of biology. An understanding of fundamental quantum interactions is important because they determine the properties of the next level of organization in biological systems. Currently, there exist four major life processes that have been identified as influenced by quantum effects: enzyme catalysis, sensory processes, energy transference, and information encoding. Quantum tunneling is of particular interest, as an explanation of the kinetics of existing reactions that should not be possible, due to insufficient energy.

Many biological processes involve the conversion of energy into forms that are usable for chemical transformations, and are quantum mechanical in nature. Such processes involve chemical reactions, light absorption, formation of excited electronic states, transfer of excitation energy, and the transfer of electrons and protons (hydrogen ions) in chemical processes, such as photosynthesis, olfaction and cellular respiration. Moreover, quantum biology may use computations to model biological interactions in light of quantum mechanical effects. Quantum biology is concerned with the influence of non-trivial quantum phenomena, which can be explained by reducing the biological process to fundamental physics, although these effects are difficult to study and can be speculative.

We could not use the same instrument (model) for all magnitudes, just same we could not use the same instruments we fix our harvester to fix our bike and watch. We need different geometry. That is why it is still difficult to integrate systematic anatomy and topographic anatomy. In physics it is even more difficult. Numerous models exist for each scale, but there should be also a universal model, that remains invariant for the whole system and does not change.

Biological rhythms, such as the circadian rhythm, are governed by molecular mechanisms within cells. The suprachiasmatic nucleus (SCN) in the brain acts as the master clock, synchronizing with environmental cues like light to regulate daily cycles of activity and rest. These biological clocks illustrate how humans are inherently time-structured at the molecular and physiological levels, with various cycles dictating behavior, metabolism, and overall health.

What is not discussed so often is the human in time dimension, regardless the fact that the same person changes predictably and dramatically over time. The building "star dust" changes insignificantly for the duration of the human life (timescales vastly longer than human lifespans or the age of the universe). The turnover of water molecules occurs on the scale of 1-2 weeks; gut lining cells: renew every 2-5 days; skin cells: renew every 2-4 weeks; red blood cells: replaced every 120 days; bone cells: turnover takes about 10 years. Carbon atoms in the body are replaced in 5-10 years due to metabolic activity. All atoms are constantly exchanged through metabolism, respiration, and digestion, and we only possess them for a short moment. The body design however remains constant and inevitable aging. So what actually changes- our perception and choice, presented by identifications.

6 Phycology

A human perception of time is a product of neural processes in the brain. Neurons in the brain's timekeeping structures, such as the basal ganglia and the cerebellum, help us perceive and measure time intervals. Cognitive neuroscience explores how the brain integrates sensory information to create a coherent experience of the passage of time. Next our identifications are developed as an inertia from past choices.

Cellular, tissue, organ and system level of organization could not be referred as a conscious structures, thus this stage of human development is mostly predetermined and following the common physics and biology laws. The zygote is the first cell, and all other emerge from it dividing itself to parts. We gather more atoms and molecules in the process, but we loss potential. By realizing choices we lose possible future conditions, with ultimately one at the end. From the, apparatus, organism and social levels of human organization, new functions emerge that enhance the human to become a subject. In philosophy, a subject is a being that exercises agency, undergoes conscious experiences, is situated in relation to other things that exist outside itself; thus, a subject is any individual, person, or observer. An object is any of the things observed or experienced by a subject, which may even include other beings (thus, from their own points of view: other subjects).

From the one cell stage of life, to 36 trillion cells stage the human changes role and reduces their potential by expanding some of all the possible trajectories. This way he builds experience and inertia for the next transformation. In this process a person becomes from more objective and fast processes to more subjective and slow processes, with the opposite tendency in the second half of the life. The reduction of the speed of life, changes the perception of time to the point of timeless experience. While the physics and biology leave no space for choosing, the psychology could built a consciousness that is self aware, time aware and able to influence biological and even underling quantum process. Through agency, the sub-personal forms of organizations are "raised" to a higher level, resulting in the personal level of experience, which is capable of overcoming the deterministic conditions and able to build coherent, unified, self-reflective and linear time illusion (Liu, Z. 2024 a).

7   Socio-culture models

Different cultures perceive and value time differently which shape how individuals experience and manage time. Some cultures have a linear view of time, emphasizing schedules and punctuality, while others may have a more cyclical, event-oriented or person-oriented perception of time. There is a great variety in social constructs of time perception, which is a proof of the existence of free will.

The examples that I have chosen to present are the Bulgarian calendar, Chinese Book of Change (I Ching), Fractal theory (chaos) and M-Theory. Fig. 2

Time	Bulgarian calendar	I Ching	Fractal theory	M-Theory
origin	divine	chaos	chaos	Quantum dimension
Flow pattern	base 60 positional numeral system	0-2-4-8-64/0/64-8-4-2-0	Cantor (ternary) set. Example of a finite subdivision rule.	11 strings vibration
Function type	Discrete (zero day)	Discrete (Primordial chaos)	Discrete (Plank distance)	Constant
System type	closed	closed	closed	closed
Free will position	Zero day	Middle state, between two predetermined sets	Quantum uncertainty	Quantum uncertainty
Compatibilism	yes	yes	yes	yes

Fig.2. Examples of very different socio-cultural frames for time measurement, that allow the presence of free will.

Very different approaches, separated by thousands of year and generations, traditions and philosophies, but with similar outcome- there is a space left for free will.

8   Bulgarian calendar

Fractal geometry serves as a bridge between the rigid geometric order of Euclid and the chaotic nature of general mathematics. Before a system's bifurcations (transformations) are fully realized, its fractal patterns may resemble a maze or puzzle. Fractals are infinitely intricate patterns that exhibit self-similarity across various scales. Chaos is filled with surprises, nonlinearities, and unpredictability. Acknowledging the chaotic, fractal nature of our world can offer new perspectives, strength, and wisdom. Once fully formed, these patterns may seem entirely determined and predictable, unless we view them as discrete functions. The monad at the start of the I Ching and Day Zero in the Bulgarian calendar both point to a system that is discrete (discontinuous) and chaotic. This suggests the necessity to seek patterns, rather than merely creating plans. The way time is divided in calendar systems has a fractal origin, evident in the infinity and self-similarity of its various cycles. In the example of the human society, this means that it should start with one person at the beginning and also end with one.

9 Ancient Chinese concept of zero and modern Fractal model

I Ching symbols are representation of a fractal system in the context of the ancient China. In this theories, integration between the deterministic time, occupying the visible part of the discrete function, and free will, occupying the transitions- the beginning (0), middle ($\varnothing$) and end parts of the function, is possible. All states and pathways through the pattern are potentially possible, but human choice navigates to one of them per each moment. Fractal model visualize that geometrically and mathematically, without any contradiction. Plank space, Zero day, Primordial chaos and Quantum effects are still beyond our full comprehension, but we experience freedom and its presence could not be scientifically excluded yet.

10 String theories

M-theory primarily addresses the physical underpinnings of the universe, not the nature of free will. However, its implications for determinism, quantum uncertainty, and the structure of reality influence philosophical discussions about human agency. Free will in this context is often considered an emergent or higher-level phenomenon, shaped by the interplay of complex systems rather than directly governed by the fundamental entities described in Mtheory. What we perceive as solid objects and empty space are composed of the same underlying strings. Human beings are essentially complex vibrational patterns of strings. This unification suggests that the same fundamental principles govern all aspects of existence, and implies a deep interconnectedness of everything in the universe. Subdivision of string coupling constant (analogy in cytology: human zygote), leads to: maximum variability, allowing freedom, allowing for free choice. The properties of the strings can be described by continuous variables, but the interactions and states they occupy are quantized.

11 Discussion

The results explore the complexity of human existence from a scientific, philosophical, and psychological perspective, particularly in relation to time. My effort is to highlight how time manifests in various dimensions-biological, psychological, sociocultural, and even quantum

– and how these dimensions interact with each other to shape human experiences. There is an ongoing lack of comprehensive understanding of the human experience of time within the larger scientific discourse. Despite the detailed exploration of how time influences our biology, behavior, and even our cultural practices, what remains underexplored is the essence of "the human" as a subject within this temporal flow. Essentially, while we understand the mechanisms of time and how it affects us, the philosophical implications of human agency within time-such as free will, consciousness, and personal identity-remain more elusive. My intention is to reduce this gap in understanding. While we can measure biological processes, time's influence on our psychology and our personal experience remains less tangible, and the role of human choice and perception in this process is still debated. Free will, as mentioned in the text, is one such area where humans' interaction with time and the universe is not fully understood. The human dimension in time is an unresolved mystery: How do we as individuals navigate, perceive, and shape time through our consciousness and choices, within the constraints of biological and physical laws? The discussion surrounding this is mainly about subjective, lived time – how it is experienced and internalized by the human mind, and how that shapes our sense of self, our agency, and our place in the universe.

If at the beginning of the 20th century there were any doubts, at the beginning of the 21st century, it was already clear that when it comes to revealing the true nature of reality, human experience is deceptive. (Greene, Brian. The hidden reality: Parallel universes and the deep laws of the cosmos. Vintage, 2011.) Since the Enlightenment, humanity has relied on science to explore fundamental questions about our identity, origins, and future. However, this endeavor has often been hindered by the assumption that we can understand the universe from an external, detached perspective. By focusing solely on external physical phenomena from this imagined viewpoint, the crucial role of subjective experience is overlooked. This limitation, termed the "Blind Spot," underpins many scientific challenges, including those related to time, the universe's origin, quantum physics, life, artificial intelligence, consciousness, and Earth's role as a planetary system. Sheldrake and his colleagues (2023) advocates for a different perspective, suggesting that scientific knowledge is an evolving, selfcorrecting narrative shaped by the interplay between the world and human experience. The argument is compelling, highlighting that all scientific experiments and inquiries are inherently influenced by the observer, making the observer an integral part of the system being studied. This approach challenges traditional assumptions, offering a refreshing perspective that encourages critical examination of foundational scientific premises.

Neurobiology gives us some ideas of the perception of time and its illusions. How are the signals entering various brain regions at varied times coordinated with one another? How aredurations, simultaneity, and temporal order coded differently in the brain? How does the brain recalibrate its time perception on the fly? All these questions could be explained with the same mechanism – by building and illusions (Eagleman, 2008). Time and space are interrelated and the flow of time can change depending on the observer's conditions (Eagleman, 2008). Mind separates time and space and creates a feeling that it flows uniformly for all, among other temporal illusions. Subjective time is a separate entity not directly dependent on physical time, though it may be associated with it via the intensity of the subject's mental activity. (Sergin, 2022). Integrated Information Theory (IIT), posits that consciousness arises from the integration of information within a system.

IIT suggests that consciousness depends on the ability of a system to combine and process information in a unified way. The theory measures the "integrated information" (denoted as Φ) of a system, with higher levels of integration corresponding to richer conscious experiences (Tononi, 2012).

The technological advances allowed us to reach beyond many of the limitations we originally have as biology structures, but this process has limits. The present crisis in physics is due to reaching such a limit. Are we going to overcome it and when is uncertain. To reduce the number of assumptions in science to only one, it should be the concept of the free will. That is required for the human life to be defined. The "blind spots" we have in our senses is also present in our logic, and it often leads our intuition in dead ends and illusions. However if the illusionary version of the universe is purposeful for supporting and reproducing life, than it is beneficial and although still an illusion, more true than the rest of the version. If all we have are illusions, at least we could choose what we need it to be. And it needs to have a "blind spot", in order to be free to choose, and to allow human existence. This type of illusion is true, because it is a center of the maximal complexity structure in the system. This changes what is the proximal and what is the distal function, and closes the system as a loop through its center, allowing for infinity to exist.

Viewing humans as a time structure in physics involves integrating insights from relativity, quantum mechanics, and cognitive neuroscience. While physical theories suggest that time is relative, non-linear, and potentially emergent, human experience and biological processes are deeply intertwined with linear perception of time. This interplay between physical theories and human perception highlights the complexity of understanding time in the context of human existence. Time is not an abstract sequence but a flow rooted in lived experience with center in our body, perception and experience. Past, present, and future are interwoven, not separate moments, creating a continuous temporal experience, through the subject's engagement with the world. Time is integral to how humans construct meaning and narratives (Merleau-Ponty, 2012).

The block universe theory, derived from the concept of spacetime in relativity, suggests that past, present, and future all coexist in a four-dimensional structure. From this perspective, humans are "time worms" extending through the block universe, with their entire lifespan existing simultaneously. This challenges the conventional view of time as a flowing entity and raises questions about free will and determinism. Despite the physical theories that describe time as a fundamental and intrinsic property, human experience of time is linear and sequential. This subjective experience is crucial for memory, anticipation, and decisionmaking.

Spiral theory of the human body states that human could be described as an auto wave; positive and negative space in the human body are composed of the same underling processes and equally important. Inspiring new approach for describing the human body and features is the use of novel mathematical tools as the Tensor analysis. It is already used to describe different parts of the human. Some authors concentrate on the movements, other on the development or the function. What is missing is a frame or a standard which agrees on the scalar base of the tensors, the number of the vectors and the structure of the matrixes used. The Spiral Theory of the Human body proposes that the M-theory could be used for a base for creating such a standard. Also it predicts that it is suitable for the individual functions to be described by a Fourier spectrum analysis, because of their spiral properties. Fourier amplitude spectrum is appropriate to be used for creation of a frequency model (pattern) of the human functions (vectors). Not only we have a short time to formulate that, but also we need to leave some time to experience a life based on it. Fourier amplitude spectrum already done for the electro-cardiogram of the human heart.

12   Conclusions

Humans are natural part of the universe and with their removal from the description, a valuable properties could not be explained. The human existence in time, should be considered as and interval, always present in the spacetime of the universe. Free will in this context is often considered an emergent or higher-level phenomenon, shaped by the interplay of complex systems rather than directly governed by the fundamental entities described in M-theory. It is reaching the maximum of complexity and becomes chaotic in nature.

Humans, as time structures, are influenced by a complex interplay of physical laws, biological rhythms, psychological perceptions, cognitive processes, and sociocultural factors. All dimensions should be studied separately for clarity, but should be interrelated if we need to have a human being.

The psychology provides us with many different questionnaires to evaluate the mental condition which vary in the number of tested categories. However the basic psychiatry interview adds several areas of interest that could be considered as psychological dimensions. Psychology could built a consciousness that is self-aware, time aware and able to influence biological and even underling quantum process. This could be the substrate of free will.

A person's life can be viewed as a line (trajectory) in four-dimensional spacetime, determined by their actions, movements and choices. The human perception of time has a beginning (birth), middle (time median) and end point (death), tracing a unique trajectory in the infinity. The human body does not exist in isolation – it is connected to its environment through gravity, motion and energy flows.

There is also the social dimension. The monad at the start of the I Ching and Day Zero in the Bulgarian calendar both point to a system that is discrete (discontinuous) and chaotic (fractal). This suggests the necessity to seek patterns. Once fully formed, these patterns may seem entirely determined and predictable, unless we view them as discrete functions.

Inspiring new approach for describing the human body and features is the use of novel mathematical tools as the Tensor analysis and Fourier amplitude spectrum.

References

Capra, Fritjof. The Tao of Physics: An Exploration of the Parallels Between Modern Physics and Eastern Mysticism. Shambhala Publications, 1999.

Eagleman, D. M. (2008). Human time perception and its illusions. Current opinion in neurobiology, 18(2), 131-136.

Greene, Brian. The hidden reality: Parallel universes and the deep laws of the cosmos. Vintage, 2011.

Grondin, Simon. "Timing and Time Perception: A Review of Recent Behavioral and Neuroscience Findings and Theoretical Directions." Attention, Perception, & Psychophysics, vol. 72, no. 3, 2010, pp. 561-582.

Hofstede, Geert. Culture's Consequences: International Differences in Work-Related Values. Sage Publications, 1980.

Levine, Robert. A Geography of Time: The Temporal Misadventures of a Social Psychologist. Basic Books, 1997.

Liu, Z. How agency is constitutive of phenomenal consciousness: pushing the first and third-personal approaches to their limits. Phenom Cogn Sci (2024). https://doi.org/10.1007/s11097-024-09968-9

Liu, Z. (2024). Husserl-Archiv der Universitat zu Koln. The 7th Spacetime conference, 2024, Albena, Bulgaria

Merleau-Ponty, Maurice. Phenomenology of Perception. Translated by Donald A. Landes, Routledge, 2012.

Sergin V.Y. Subjective Time: Nature and Neurobiological Mechanisms. Neurosci Behav Physi 52, 1082-1092 (2022). https://doi.org/10.1007/s11055-022-01336-x)

Sheldrake, Rupert. The Blind Spot: Why Science Cannot Ignore Human Experience. Counterpoint Press, 2023.

Tononi, Giulio. Phi: A Voyage from the Brain to the Soul. Pantheon Books, 2012.

Wirth, M., Wettstein, M., & Rothermund, K. (2024). Longitudinal associations between time perspective and life satisfaction across adulthood. Communications Psychology, 2(1), 67.

Part IV

PASSAGE AND DIRECTION OF TIME

Kyley Ewing (Ed.), *Spacetime Conference 2024. Selected peer-reviewed papers presented at the Seventh International Conference on the Nature and Ontology of Spacetime, 16 - 19 September 2024, Albena, Bulgaria* (Minkowski Institute Press, Montreal 2025). ISBN 978-1-998902-44-6 (softcover), ISBN 978-1-998902-45-3 (ebook).

9 THE HAPPENING OF EVENTS IN TEMPORAL RELATION TO ONE ANOTHER ESTABLISHES BISTABLE AND SELFSUSTAINING DYNAMICS OF THE UNIVERSE

BRUCE M. BOMAN

Abstract   My goal is to study the dynamics of the Universe from a relational perspective based on the happening of events in temporal relation to each other and their respective points of reference. Accordingly, the flow of time was modeled as the progression of events that happen at the microscale relative to the macroscale. Kinetic modeling was predicated on transition state theory, and the direction of the reaction progresses from future events to now events to past events whereby past events also auto-catalyze their own production. Model rate equations and rate laws were used to analyze if the system possesses stable equilibria, steady state, perpetual, and bistable dynamics. The results show that the Universe is self-sustaining and exists as a dynamic bistability that has two distinct, stable equilibrium states. The idea that the Universe exists as a dynamic relation between two stable equilibrium states and switches between them through hysteresis might provide a means to understand new mechanisms that explain the dynamic structure of the Universe.

Introduction

My goal is to study the dynamics of the Universe from a relational perspective whereby events can only exist in relation to other events. Indeed, in his book Helgoland, Carlo Rovelli deliberates on relational quantum mechanics including what the Buddhist philosopher Nagarjuna teaches us about ideas on relativity [1]. Nagarjuna's main idea is that the determination of a thing or object is only possible in relation to other things or objects, especially when considering them by way of contrast. The classical example he gives us is that shortness exists only in relation to the idea of length. In this line of thinking, I would argue that determination of the existence of events is only possible in relation to the dynamics of the Universe. Accordingly, the research question is how events happen in relation to the happening of other events. To seek an answer, the flow of time was modeled as the progression of events.

In my modeling of the progression of events at the microscale as a kinetic reaction, the direction is time events progress from future events to now events to past events. This direction of progression of time events is similar to the progression of a chemical reaction in which the reactants interact and form an activated complex during their transition to form products. However, this direction of time events is in contrast to the notion in physics that time flows from past to present to future. Because my concept of the direction of time progression is opposite to the direction that is considered "normal" by the scientific

Kyley Ewing (Ed.), *Spacetime Conference 2024. Selected peer-reviewed papers presented at the Seventh International Conference on the Nature and Ontology of Spacetime, 16 - 19 September 2024, Albena, Bulgaria* (Minkowski Institute Press, Montreal 2025). ISBN 978-1-998902-44-6 (softcover), ISBN 978-1-998902-45-3 (ebook).

community, I will start by: 1) Discussing the views that other scientists have on the direction of time; 2) Explain why I believe that the flow of time at the macroscale occurs from past to present to future, but at the microscale, the progression of events happens from future to now to past.

Views by different scientists on the direction of time

Certainly, a vast literature exists on the direction of time. In his book, "History of Eternity" [2], Jorge Luis Borges contends that flow from past to future and the opposite direction are both equally probable and equally unverifiable. Moreover, in his famous essay, "A New Refutation of Time" Borges goes on to argue that each moment is self-contained and absolute, which challenges the traditional concept on flow of time and negates the concept of past, present, and future [3].

In metaphysics, the direction of time appears anisotropic depending on whether it is viewed from earlier to later as opposed to from later to earlier. Still this doesn't mean that time has a direction because that would require providing the existence of an objective fact as to whether or not time goes from earlier to later or later to earlier. In his perspective in "The Direction of Time" Sam Baron discusses the ongoing debates in the literature as to whether, or not, time has a direction [4]. For two of the main groups of theorists (A-theorists and B-theorists), both believe that time has an objective direction, while C-theorists deny this notion and believe that time lacks a direction [5]. Clearly, over the past decades, there is much debate among these theorists [4-8]. For A-theorists, their view is based on presentism whereby only the present is real. In contrast, the ideas held by B-theorists are usually based on eternalism whereby all events in time (past, present, future) exist equally and simultaneously. Eternalism is often equated with the "block Universe" theory that holds spacetime is a four-dimensional block that doesn't change, so there is no inherent direction in the flow of time. In fact, Julian Barbour's views on time go beyond the block Universe by proposing that the passage of time is an illusion and instead exists as a relational concept [9].

In other views on time's direction, an accounting for the existence of temporal asymmetries is problematic because the laws of nature are time reversal invariant. Indeed, at the most fundamental level, the equations of physics do not inherently contain a direction of time, which means that time itself is not flowing in a single direction. Still, the prevailing belief in physics is that time flows from past to present to future. To explain this direction of time, theorists refer to specific features of the Universe such as expansion of the Universe, increase in entropy over time, and causal order. Additionally, we actively think, plan, and make choices about things that are going to occur in the future, but we can't change what's already happened in the past. Thus, it is our perception and experience of time which usually leads us to believe time flows from past to present to future.

The traditional view that the direction of time flow from the past into the future means that time flows from past events that occurred earlier to future events that will occur later. In this view, the flow of time as moving from "earlier" to "later" also signifies that the past is different from the future, and that time only moves forward and never backwards. Such a view is often described as "one-way direction" or "asymmetry" of time [8]. This asymmetry has also been described as the increase in entropy over time. In thermodynamics, the arrow of time has been related to the second law of thermodynamics, which states that in isolated systems, entropy tends to increase with time.

Many physicists, including famous ones like Albert Einstein, Arthur Eddington, and Stephen Hawking, contributed to our understanding of the direction of time by linking the "arrow of time" to increasing entropy and the expansion of the Universe [10]. For example, Steven Hawking studied the cosmological arrow of time, relating it to the expansion of the Universe and the "no boundary proposal" which indicates the beginning of the Universe occurred in a smooth, ordered state and then evolved towards disorder [11].

Moreover, the theoretical physicist Carlo Rovelli, ascribed the "arrow of time," as arising from the concept of entropy. Indeed, in his book [12], The Order of Time, he states that "It's entropy not energy that drives the world". In his 2018 talk entitled "The Physics and Philosophy of Time" [13], he claims that "There is nothing else in the world that distinguishes the past from the future except this entropy".

Additionally, the theoretical physicist Sean Carroll studied the "arrow of time" in relation to entropy and cosmology claiming that the low entropy past of the Universe is a reasonable source for the direction of time [14, 15]. He goes on to claim that the second law of thermodynamics, which stipulates entropy in an isolated system tends to increase, is key to understanding the arrow of time. The Nobel Laureate Roger Penrose claimed that entropy and second law of thermodynamics are fundamental to understanding the direction of time [16].

Overall, many physicists will fall into one of two camps: 1). The "entropy camp" which believes the arrow of time is explained by the increasing entropy and the initial low entropy state of the Universe, and 2). The "extrinsic camp" which suggests the arrow of time is an inherent feature of the Universe, requiring a new fundamental law [17].

More recently, the theoretical physicist Lee Smolin argued that, even though the laws of physics themselves seem to be time-symmetric, the arrow of time is a fundamental asymmetry that needs explanation. While the increase in entropy (disorder) in the Universe is often cited as a factor contributing to the arrow of time, Smolin suggested that this explanation alone is not sufficient. [18].

Consequently, the notion that the "arrow of time," arises from the concept of entropy needs a closer look. Actually, the physical chemist Arieh Ben-Naim argues that there is no inherent connection between entropy and flow of time [19,20]. He claims that the common association between time's arrow and entropy arises due to misinterpretations of the Second Law of Thermodynamics. In his perspective, entropy is a static concept that describes the number of possible microstates that a system can exist in. Thus, the concept of entropy itself does not dictate the arrow of time and doesn't inherently point to a direction of time flow. He even goes on and challenges Boltzmann's H-Theorem as providing evidence for the arrow of time. This problem arose because the theory is misunderstood as relating to entropy when it actually relates to information content. Therefore, our understanding of entropy should be based on a statistical mechanics perspective, which focuses on probabilities of various microstates, rather than an overly simplistic and misleading interpretation based on disorder.

Given that the mystery to the arrow of time is yet to be solved, I began modeling the progression of time events based on reaction kinetics. Before I go over the details in my model, I will present the initial approach I took in designing a model.

Approach

My approach to develop a model for the kinetics of events, was to understand how events happen. For this task, it was important to distinguish the difference between why events happen versus how events happen. The difference is: "why" asks about the reason or cause behind something, while "how" asks about the process or method used to achieve something. Essentially, "why" focuses on the purpose, while "how" focuses on the steps involved in an action. Definitions [21] for why and how are given below.

> Why = for what reason or purpose. Example, why did you drive to the store? I was out of milk and eggs so I drove to the store.
> How = in what manner or way, by what means. Example, how did you drive to the store? I got in my car, drove for 8 minutes, and then arrived at the store.

As I will discuss in the next section below, when considering "why things happen" as compared to "how things happen", two different interpretations are elucidated for the progression of events at the macroscale versus the microscale.

For the purpose of discussion on the progression of events, the following terms are provided for future, now, and past events.

> Future event = yet to happen
> Now event = is happening
> Past event= has happened.

Using this approach, I began modeling the progression of time events based on reaction kinetics. In the next section, I will present my thinking on the flow of time from past to present to future at the macroscale. Then, I will explain why I believe that the progression of events occurs from future events to now events to past events at the microscale.

PROGRESSION OF EVENTS AT THE MACROSCALE VERSUS MICROSCALE.

The flow of time is often described as a continued sequence of events that succeed one another [22]. Based on this view, the flow of time can be described as a progression of events which occur in irreversible succession. Indeed, Carlo Rovelli states in the "The Order of Time", that "Ours is a world of events not things" [12]. The Merriam-Webster Dictionary defines event as "Something that happens" [23]. In this sense, it is actually the happening of events, rather than the flow of time, that is fundamental [24]. Thus, events don't happen in spacetime, rather spacetime is the happening of events. Thus, in my study, spacetime is the happening of events, and the happening of events is produced by an interaction of events with each other.

Progression of events at the macroscale

When the progression of events is perceived at the macroscopic scale, this happens in relation to the occurrence of other events that happened over time. For example, from the human-centric perspective, as we age we can remember more and more events that happened over our lifetime. We can plot this as a graph in terms of cumulative number of events relative to our age (Figure 1). We can also do this for the global scale since the time of the presumed big bang (Figure 1). Since temporal relations are crucial for defining why things happen based on when they happen and for understanding cause-and-effect scenarios, in Figure 1, the direction of progression of events happens from past to now to future. In this macroscopic view, now events are caused by events that occurred before them, and then these causes lead to specific consequences that shape future events [25,26]. The classic example of this causality is the light cone that illustrates the evolution of a flash of light emanating from a single event and traveling in all directions in Minkowski spacetime [27]. The future light cone contains all the events that can potentially be causally influenced by the initial event. The past light cone contains all the events that could possibly influence the initial event. The big problem is that there is no such thing as a universal reference frame, so it is only possible to gauge events in relation to something else (i.e. other events). This makes it really difficult to determine the "now, past, and future" at the macroscale except in relation to events that have already happened.

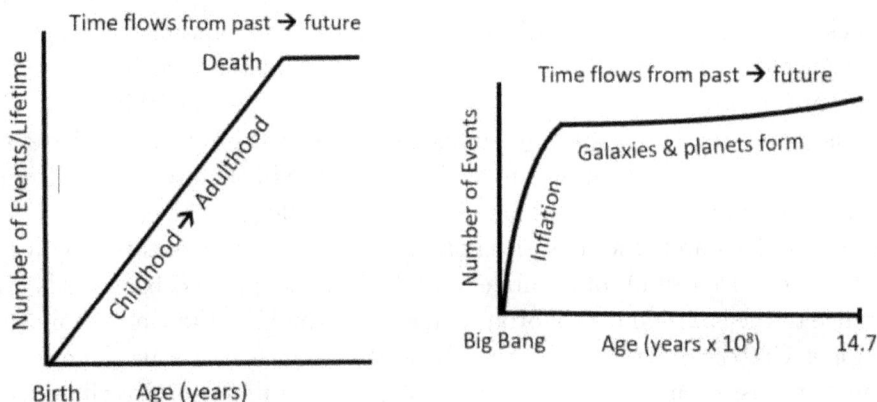

Figure 1. Cumulative number of events relative to our age (Left Panel), and at the global scale since the time of the big bang (Right Panel).

Progression of events at the microscale

At the microscale, I consider how events progress locally at each point of reference. In this perspective, there is a before, during, and after the happening of an event, i.e. 1) a future event is yet to happen, 2) a now event is happening, and 3) a past event has already happened. There are many examples in the physical sciences that have this perspective on reactions, which are used to develop kinetic models.

In thermodynamics and statistical mechanics, Boltzmann's equation has been used to describe the distribution of kinetic energies among particles in a system which is not in a state of equilibrium. So, at a given temperature, some particles will have more energy than others, and a higher proportion of them will have sufficient energy to react when they collide. For example, a fluid with temperature gradients in space will cause conduction of heat from hotter regions to colder ones by the transport of the particles that makes up the fluid. Because the particles interact through collision, each collision will occur at a point of reference and the system will have an infinite number of points of reference. Thus, at the microscale, this process by which heat or energy is conducted from a difference in temperature is actually a kinetic reaction involving the collision of particles. Therefore, at each point of reference there will be an event yet to happen, and event that is happening, and an event that happened.

In theoretical physics, Feynman diagrams represent mathematical expressions that describe the behavior and interaction of subatomic particles. One of Feynman's classic diagrams depicts the interaction of an electron and a positron which annihilate to produce a photon that becomes a quark-antiquark pair after which the antiquark generates a gluon. Thus, at each point of reference there will be an event yet to happen, and event that is happening, and an event that happened.

Moreover, in his Theory of Positrons [28], Feynman's solution to the problem of the behavior of positrons and electrons was given as "negative energy states" that appear in a form which may be pictured in spacetime as waves traveling away from the external potential backwards in time. He describes the wave as a positron approaching the potential and annihilating the electron. So, the electron moving forward in time in a potential may be scattered forward in time (ordinary scattering) or backward (pair annihilation). Thus, when the positron is moving backward, it may be scattered backward in time (positron scattering) or forward (pair production). Consequently, the amplitude in the progression from an initial to a final state can be analyzed by considering that it undergoes a sequence of scatterings. As a result, the actions and reactions of particles, when viewed as events, might give rise to the notion that the progression of events can occur in forwards and backwards directions.

In reference to the collision of a photon with an electron [29], the German mathematician and philosopher Grete Hermann noted the physical system of the wave function is a linear combination of terms, each being "the product of one wave function describing the electron and one describing the light quantum". She goes on to state "The light quantum and the electron are thus not described each by itself, but only in their relation to each other. Each state of the one is associated with one of the other."

Recently, researchers conducted experiments to study how photons interact with atoms in a medium by using a cloud of cold rubidium-85 atoms that was illuminated by a pulse of resonant light [30]. They observed that the atoms were excited before the light even arrived. One interpretation of this finding is that photons may cause "negative" time by exciting atoms as they pass through the medium. Even though this interpretation is still being debated, this study shows that interactions in quantum systems can only be described from a relational perspective based on the happening of events in relation to one another.

Another recent study involved investigation on the mathematical correlations between the behavior of qubits and the geometry of space [31]. The study shows that time is best understood as the succession of present moments that happen one after another. This result on temporal correlations also suggests that space itself may even emerge from the structure of quantum temporal correlations. Moreover, this study suggests that the Universe needs to be defined from a relational perspective whereby space can only exist in relation to temporal correlations. In the present study on spacetime, I explore this notion as the progression of events.

In studying the progression of events, I model it as a kinetic reaction like a chemical reaction. For chemical reactions, the reactants are in continuous motion and constantly interacting with each other at each and every reference point. Hence, kinetic molecular theory, collision theory and rate equations are used to model the kinetic activity of the reactants by describing their behavior at the microscopic level, including random motion, collisions, and energy transfer. Once a rate equation is established, it is used to predict how the rate of the reaction changes under different conditions, including changes in concentrations of reactants and presence of a catalyst. In a chemical reaction, the constituent atoms of the reactants are mutable as they interact and rearrange to create different substances as products. In spontaneous chemical reactions, such as combustion, the products are immutable because they form irreversibly. Thus, the happening of now events can be viewed as spontaneous reactions that occur naturally because past events are formed irreversibly and are immutable.

Taken together, each of the above examples illustrates that the progression of events can be viewed as a kinetic reaction. When viewed at the microscale, the happening of an events at each point of reference is preceded by an event yet to happen and followed by an event that already happened. In my model, I consider these events as future events, now events, and past events from the perspective of the progression of events at a point of reference. Next I discuss the progression of events from a relational perspective based on the happening of events in temporal relation to each other and their respective points of reference.

A relational perspective on the happening of events

To understand the progression of events at the microscale, I analyzed how the progression of events occurs from future to now to past based on how events happen locally, not globally. The ordering of events at the microscale is illustrated by using the 8 minute Sun to Earth example. In this analysis, it's essential to define the "when" and "where" at different points of reference. In the diagram (Figure 2), there are two points of reference: POR 1) the Sun, and POR 2) the Earth. A time now event is depicted to happen at each point of reference: POR 1) the generation of light from nuclear fusion on the Sun, and POR 2) reflection of the sunlight on the Earth. Once a time now event happens at each reference point, it then instantaneously becomes a time past event at that reference point. A remarkable example of reflection of light from Earth is the photograph of Earth taken by NASA's Voyager 1 Spacecraft when it was 6 billion kilometers from Earth. This image of a tiny speck in space was called the "Pale Blue Dot" by Carl Sagan in his book "A Vision of the Human Future in Space" [32]. From a relational perspective, this reflected light is a past event from the Earth's reference point, but it is a future event from Voyager 1's reference point.

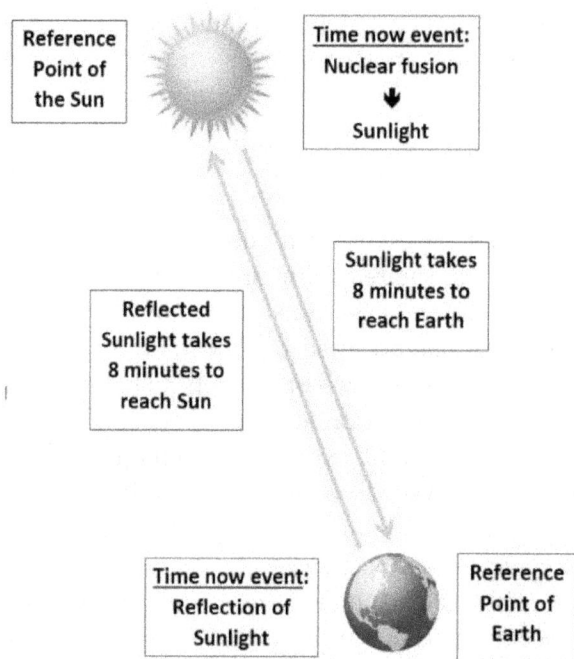

Figure 2. Schematic shows two points of reference: *POR* 1) the Sun, and *POR* 2) the Earth. It takes 8 minutes for light to travel from the Sun (*POR 1*)to the Earth (*POR 2*) and vice versa.

For events happening on Earth, there are innumerable "now" events that are happening when light from the Sun reaches the Earth, which then irreversibly progress to past events. For example, light reaching the retina in the eye can trigger a neural signal, photons interacting with a solar panel can energize electrons in generating electricity, and the chlorophyll in plants uses sunlight to create energy-carrying ATP molecules. In each example, the progression to a past event is irreversible.

The following paragraph provides a temporal relation perspective on events that occur at these different reference points shown in Figure 2. During the 8 minutes it takes for the sunlight to reach the Earth from the Sun, the sunlight is considered to be: *i*) a past event that happened at the original reference point of the Sun, and ii) a future event that is yet to happen at the reference point of the Earth. Similarly, during the 8 minutes it takes for reflected sunlight to reach the Sun from the Earth, the reflected sunlight can considered to be: *i*) a past event that happened at the reference point of the Earth and *ii*) a future event that is yet to happen at the reference point of the Sun. In each example, the sunlight is a past event at one point of reference and a future event at the other point of reference based on the order of how events happen.

In this scenario, a past event only happens after, not before, the moment when the sunlight progresses as a now event to a past event locally at the respective point of reference. In this thinking, when the sunlight reaches the Earth and when the reflected sunlight reaches the Sun, both these events are now events from their respective points of view. Thus, both of these now events are defined according to the timing and order that events are happening locally.

Based on this line of reasoning, during the 8 minutes it takes for sunlight to reach Earth, the sunlight can be considered to be a future event that hasn't happened yet on Earth. Similarly, the reflected sunlight can be considered to be a future event that hasn't happened yet on the Sun. Thus, when considering events at one point of reference in relation to events at another point of reference, it illustrates that past events happening at one point of reference can be future events that will happen at another point of reference.

It also illustrates that there could be a reversible-type of dynamic (backward and forward relation) between events that are happening at different points of reference relative to other points of reference. This might even be thought of as an equilibrium involving a forward reaction and backward reaction between future events and now events. Since sunlight generated from our Sun and other stars can travel throughout the Universe, it is worthwhile to consider whether this relational dynamic could exist globally between every event happening at all points of reference. Clearly, events are continuously in motion and interacting throughout the Universe that gives structure to the Universe.

In summary, when events are viewed at the microscale based on a relational perspective whereby events happen in relation to other events at different points of reference, the results indicate that the direction of time differs from the conventional view at the macroscale. Based on the above deductive reasoning, it is hypothesized that the progression of events at each point of reference occurs from future events to now events to past events. This progression of events at the microscale also fits well with our understanding of mechanisms involved in kinetic reactions that progress from reactants to transition state to products. For example, the past is an irreversible event formed similar to the products that are irreversibly formed in a kinetic reaction. Moreover, the transition state in a kinetic reaction occurs instantaneously which is analogous to the present which happens only momentarily. Additionally, in kinetic reactions, reactants are in continuous motion and interacting in the transition to products, which is similar to future events that are mutable by continuously interacting with other events. Accordingly, herein, I have modeled the progression of time events based on concepts from transition state theory of kinetic reactions in chemistry to understand the dynamic nature of events in our micro-world.

Modeling the Progression of Time Based on Transition State Theory

Background

Transition State Theory (TST) provides a mechanism-based approach to understand the dynamics of time and space (in terms of reaction rates) from a reaction kinetics perspective [32-36]. Modeling that time progresses because of the interaction of time events at a given point of reference lets us understand how the progression of time events and their asymmetric direction occur as a spontaneous reaction based on reasoning from TST (Figure 3). This theory has been used for nearly 100 years to understand mechanisms of kinetic reactions. In TST, the interaction of the reactants leads to formation of an instantaneously activated complex during a transition state in the progression to products [37-39]. TST also assumes that a quasi-equilibrium exists between the reactants and the activated complex in the transition state. The energy of activation (E_a) is a threshold energy that the reactant(s) must acquire before reaching the transition state. Once in the transition state, the reaction can go in the forward direction towards product(s), or in the opposite direction towards reactant(s) [40,41]. The reaction products are then irreversibly formed from the transition state. Thus, the level of E_a that a reaction requires to reach the transition state controls rate of progression to reaction products. By applying TST to explain progression of events in spacetime, we can begin to understand the thermodynamics (energy & entropy) in the transition of future time events to past time events.

TST also provides a mechanism-based approach to understand why the progression of time events might be perpetual. In modeling the progression of time, the model mechanism also involves an interaction of past time events with time now events, which catalyzes the production of past time events. That is, past time events catalyze their own production. In chemistry, a reaction is autocatalytic if one of the reaction products is also a catalyst for the same reaction. In this view, the autocatalytic production of past time events may provide a mechanism that explains why the progression of time is self-sustaining.

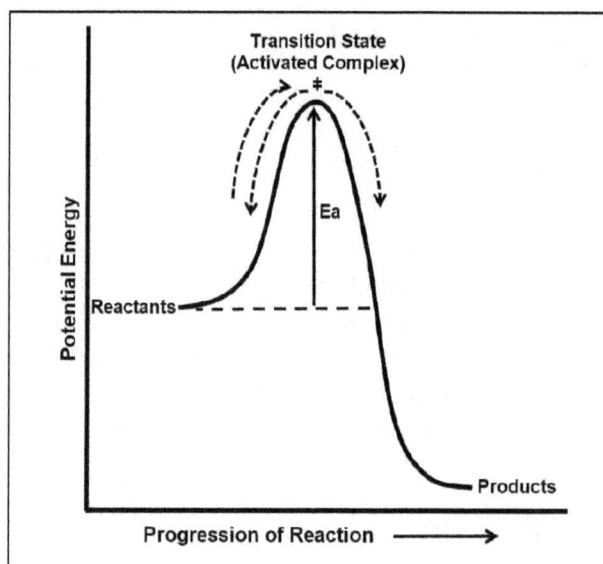

Figure 3. **Potential energy diagram for the progression of a chemical reaction.** The progression of a chemical reaction is explained by TST in terms of activation energy (E_a) and enthalpy of formation (Δ_H). The progress of the reaction is from reactants to transition state to products. The highest position for energy of activation (E_a) required to initiate the reaction represents the transition state. The energy difference between the reactants and transition state equals the activation energy of the reaction. This energy is directly related to the rate of the reaction, which is determined by the shape of the potential energy curve connecting the reactants and products. The transition state (activated complex) is an instantaneous point in the conversion of reactants to products. Also note, in TST, it is assumed that an equilibrium exists between the reactants and the transition state-activated complex.

MODEL DESIGN

Accordingly, a mathematical model was created for the kinetics of spacetime events whereby an activated transition state occurs as a now time event in the progression of future time events to past time events. The current model builds on our previous modeling of the progression of time based on TST [42-46]. Here, a new autocatalytic reaction step is introduced whereby time past (t_P) events interact with time now (t_N) events which leads to an activated t_N^* event. In this way, the present (time now) is considered to be an instantaneous transition state in the progression of future to past. Additionally, based on TST, a quasi-equilibrium exists between future time (t_F) events and now time (t_N) events that establishes a superposition-like state whereby time fluctuates forward and backward between future time and now time. From the transition state, the activated time now (t_N^*) events will irreversibly progress to past time (t_P) events. Past time (t_P) events are different from future time (t_F) and now time (t_N) events because past time (t_P) events are immutable. Still, time past (t_P) events can interact with time now (t_N) events in an autocatalytic reaction mechanism to drive the perpetual dynamics of time event progression.

Definition of an Event

In my model, spacetime is not defined in terms of fixed spatial or temporal dimensions, but simply as a dynamic variable (an event) based on changes in time as a function of changes in space (a spacetime vector or tensor). As noted above, this is consonant with the real world as our physical measurement of time is based on changes in space. Thus, an event is defined as any change that takes place involving both time and space. And, the progression of spacetime is measured as a sequence of interacting events in motion at a given reference point where time changes (dt) as a function of changes in space (ds). In addition, time and space are variables expressed in terms of units of time and space. In scalar terms, an event is a change in time relative to a change in space which occurs at all given points of reference in the Universe.

Accordingly, in the model, the proposed quantitative definition of an event is as follows: $\text{Event} = \frac{\text{Change in time}}{\text{Change in space}} = \frac{dt}{ds} = \frac{t_2 - t_1}{s_2 - s_1}$, where t = time and s = space.

Model Assumptions

The model assumptions on which it is based, are discussed below.

1. Time past (t_P), time now (t_N), and time future (t_F) events are distinct from one another.

2. Time now (t_N) is an instantaneous transition state between t_F and t_P in the progression of t_F to t_P Namely, time future (t_F) progresses to time now (t_N) which progresses to time past (t_P).

3. Time now (t_N) is also considered to occur at a point of reference anywhere in space where events progress in one direction and time progresses from future to past. In this

191

view, some events, termed time future (t_F), are happening prior to (precede) time now (t_N) and are coming to be in time now. Other events, termed time past (t_P), take place later (downstream) than at the given point of reference of time now (t_N).

4. A quasi-equilibrium exists between t_F and t_N events. The equilibrium exists because not every interaction of t_F events results in the formation of the transition state. Based on the equilibrium between t_F and t_N events, the progression of spacetime events occurs according to thermodynamics (changes in energy and entropy) of the system.

5. Once interacting t_F events reach a transition state, a t_N event is formed. At that instant, it can interact with t_P events to progress to an activated t_N^* event which does not collapse back to the t_N event. Time past (t_P) events arise from an activated t_N event (t_N^*) through an irreversible reaction from t_N^* to t_P.

6. The interaction of time past (t_P) events with time (t_N) events to form activated time now (t_N^*) events that progress to time past (t_P) events simulates the dynamics of an autocatalytic mechanism whereby time past (t_P) events catalyze their own production.

$$t_N + t_P \rightarrow t_N^* \rightarrow t_P + t_P$$

7. Time past (t_P) events then progress to time future (t_F) events. This reaction simulates the idea that t_P events at one point of reference are time future (t_F) events in relation to another point of reference.

Thus, in the model, time past (t_P) events are incorporated as reactants that catalyze their own production in order to drive the dynamics of an autocatalytic mechanism. The presence of a t_P event that catalyzes its own production is predicted to accelerate its own production. The individual reaction steps and rate equations are given below.

Individual Reaction Steps

$$t_F + t_F \xrightarrow{k_1} t_N \qquad \text{(rxn. 1.1)}$$

$$t_N \xrightarrow{k_2} t_F + t_F \qquad \text{(rxn. 1.2)}$$

$$t_N + t_P \xrightarrow{k_3} t_N^* \qquad \text{(rxn. 1.3)}$$

$$t_N^* \xrightarrow{k_4} t_P + t_P \qquad \text{(rxn. 1.4)}$$

$$t_P \xrightarrow{k_5} t_F \qquad \text{(rxn. 1.5)}$$

Rate Equations

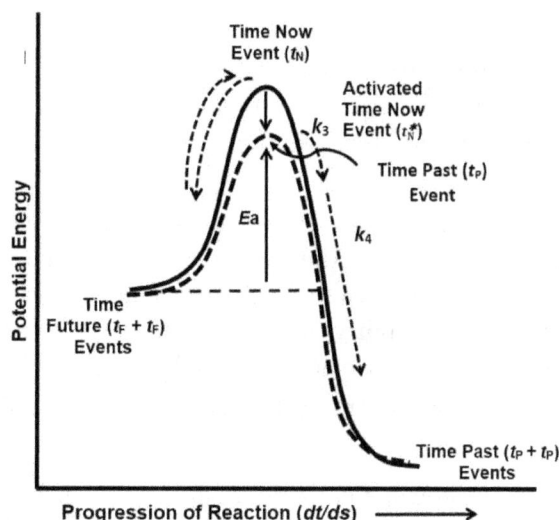

Figure 4. Potential energy diagram showing the interaction between a time past (t_P) event and a time now (t_N) event that lowers the energy of activation (E_a). In this model, the present (time now) is an instantaneous transition state in the progression of future events to past events. If the E_a is insufficient to pass the energy barrier, a quasi-equilibrium exists between future time (t_F) events and now time (t_N) events that establishes a superposition-like state whereby time fluctuates forward and backward between future time and now time. Past time (t_P) events are different from future time (t_F) and now time (t_N) events because past time (t_P) events are immutable. Still, time past (t_P) events can interact with time now (t_N) events, which leads to an activated t_N^* event that irreversibly progresses to past time (t_P) events.

$$\frac{dt_F}{ds} = 2k_2t_N - k_1t_F^2 + k_5t_P \qquad \text{(eq. 1.1)}$$

$$\frac{dt_N}{ds} = k_1t_F^2 - k_2t_N - k_3t_Nt_P \qquad \text{(eq. 1.2)}$$

$$\frac{dt_N^*}{ds} = k_3t_Nt_P - k_4t_N^* \qquad \text{(eq. 1.3)}$$

$$\frac{dt_P}{ds} = 2k_4t_N^* - k_3t_Nt_P - k_5t_P \qquad \text{(eq. 1.4)}$$

$$\frac{d\left(t_F + t_N + t_N^* + t_P\right)}{ds} = k_2t_N + k_4t_N^* - k_3t_Nt_P \qquad \text{(eq. 1.5)}$$

(Note that, for first order reactions, rate constants have units of s^{-1}; for second order reactions, the rate constant has units of $s^{-1}t^{-1}$.)

Reaction kinetics ideas and conditions studied in the model.

Equilibrium. In kinetic systems, equilibrium refers to a reversible reaction in which the rate of the forward reaction is equal to the rate of the reverse reaction. In the scenarios below, existence of two possible equilibrium states is determined by showing that $\frac{d(t_F+t_N)}{ds} = 0$ or $\frac{d(t_N^*+t_P)}{ds} = 0$.

Steady State. In kinetic systems, steady state refers to a condition where all the variables in the system remain constant even though the reaction continues to happen. In the scenarios below, existence of a possible steady state of the overall system is determined by showing that $\frac{d(t_F+t_N+t_N^*+t_P)}{ds} = 0$.

Autocatalysis. In kinetic systems, autocatalysis involves a reaction in which one of the products also functions as a catalyst for the same reaction. In the model, an autocatalytic reaction occurs as $t_N + t_P \rightarrow t_N^* \rightarrow t_P + t_P$, whereby t_P events catalyze their own production.

Bistability. In a kinetic system, bistability means that the system has two stable equilibrium states and can switch between the two states (via hysteresis). In the scenarios below, existence of a possible bistable state in the system is determined by showing that $\frac{d(t_F+t_N)}{ds} = \frac{d(t_N^*+t_P)}{ds} = 0$.

Cause and Effect. In kinetic systems, a cause-and-effect relationship occurs where a change in one event (the cause) triggers another event to occur (the effect). In the proposed kinetic reaction, a cause-and-effect relationship is modeled as the interaction of a time past (t_P) event with a time now (t_N) event. The effect of the $t_P + t_N$ interaction (the cause) is a decrease in the energy of activation that leads to an activated t_N^* event which then passes the energy barrier and irreversibly progresses to past time (t_P) events.

Three Scenarios are presented below to illustrate how perturbations in the system affect the steady state dynamics and progression of time events

SCENARIO 1. To study the effect of events having insufficient energy to pass the energy barrier, the k_3 rate constant was set to zero so that the transition from t_N to t_N^* was blocked. If E_a of interaction between two time future events ($t_F + t_F$) is not equal to or greater than the minimum required for the transition to take place (i.e. $k_3 = 0$), this leads to a reversible reaction between the forward reaction $t_F + t_F \xrightarrow{k_1} t_N$ and the backward reaction $t_N \xrightarrow{k_2} t_F + t_F$, although t_F is still produced from t_P through the rxn 1.5 $\left(t_P \xrightarrow{k_5} t_F \right)$. Below, the objective is to determine if one or more equilibrium states exist.

In this scenario, $\frac{dt_F}{ds} = 2k_2 t_N - k_1 t_F^2 + k_5 t_P$ (eq. 1.1) and $k_3 = 0$

Since $\frac{dt_N}{ds} = k_1 t_F^2 - k_2 t_N - k_3 t_N t_P$ (eq. 1.2) and $k_3 = 0$

$$\text{then } \frac{dt_N}{ds} = k_1 t_F^2 - k_2 t_N \qquad \text{(eq. 1.6)}$$

$$\text{So, } \frac{d(t_F + t_N)}{ds} = k_2 t_N + k_5 t_P \qquad \text{(eq. 1.7)}$$

Thus, when $k_3 = 0$, an equilibrium does not exist between t_F and t_N events.

Since $\dfrac{dt_N^*}{ds} = k_3 t_N t_P - k_4 t_N^*$ (eq. 1.3) & $\dfrac{dt_P}{ds} = 2k_4 t_N^* - k_3 t_N t_P - k_5 t_P$ (eq. 1.4)

when $k_3 = 0$, then $\dfrac{d\left(t_N^* + t_P\right)}{ds} = k_4 t_N^* - k_5 t_P.$ \hfill (eq. 1.8)

Thus, when $k_3 = 0$, an equilibrium can exist between t_N^* and t_P events, if $k_4 t_N^* = k_5 t_P$, so $\dfrac{d\left(t_N^* + t_P\right)}{ds} = 0$.

So, when $k_3 = 0$, based on eq. (1.5),

$$\frac{d\left(t_F + t_N + t_N^* + t_P\right)}{ds} = k_2 t_N + k_4 t_N^*. \tag{eq. 1.9}$$

Overall, in Scenario 1, where $k_3 = 0$, an equilibrium does not exist between t_F and t_N events, but, an equilibrium can exist between t_N^* and t_P events. Also, the overall system doesn't depict a steady state.

SCENARIO 2. Further analysis was done by setting $k_5 = 0$ to test the effect of blocking the reaction $t_P \xrightarrow{k_5} t_F$ (rxn 1.5). In this scenario, the E_a in the interaction between the $t_F + t_F$ events is equal to or greater than the minimum that is required for a transition to occur (i.e. $k_3 \neq 0$), but the formation of t_F events from t_P events is zero (i.e. $k_5 = 0$).

In this scenario, $\dfrac{dt_F}{ds} = 2k_2 t_N - k_1 t_F^2 + k_5 t_P$ (eq. 1.1) and

$$\frac{dt_N}{ds} = k_1\, t_F^2 - k_2 t_N - k_3\, t_N\, t_P. \tag{eq. 1.2}$$

So, when $k_5 = 0$, then

$$\frac{d\left(t_F + t_N\right)}{ds} = k_2 t_N - k_3 t_N t_P. \tag{eq. 2.0}$$

Thus, when $k_5 = 0$, an equilibrium can exist between t_F and t_N events if $k_2 t_N = k_3 t_N t_P$, so $\dfrac{d\left(t_F + t_N\right)}{ds} = 0$.

Since $\dfrac{dt_P}{ds} = 2k_4 t_N^* - k_3 t_N t_P - k_5 t_P$ (eq. 1.4), when $k_5 = 0$,

$$\frac{dt_P}{ds} = 2k_4 t_N^* - k_3 t_N t_P. \tag{eq. 2.1}$$

Since, $\dfrac{dt_N^*}{ds} = k_3 t_N t_P - k_4 t_N^*$ (eq. 1.3), thus

$$\frac{d\left(t_N^* + t_P\right)}{ds} = k_4 t_N^*. \tag{eq. 2.2}$$

Thus, when $k_5 = 0$, an equilibrium does not exist between t_N^* and t_P events.
Also, when $k_5 = 0$,

$$\frac{d\left(t_F + t_N + t_N^* + t_P\right)}{ds} = k_2 t_N + k_4 t_N^* - k_3 t_N t_P. \tag{eq. 2.3}$$

Overall, in Scenario 2 where $k_5 = 0$, an equilibrium can exist between t_F and t_N events (eq. 2.0), but an equilibrium does not exist between t_N^* and t events (eq. 2.2). But, the overall system can exist in a steady state when $k_2 t_N + k_4 t_N^* = k_3 t_N t_P$, such that $\frac{d(t_F + t_N + t_N^* + t_P)}{ds} = 0$ (see eq. 2.3).

Thus, a steady state exists if $k_2 t_N + k_4 t_N^* = k_3 t_N t_P$, and $\frac{d(t_F + t_N + t_N^* + t_P)}{ds} = 0$.

SCENARIO 3. The dynamics of the overall system were analyzed in this scenario. The E_a in the interaction between the $t_F + t_P$ events is equal to or greater than the minimum that is required for a transition to occur, i.e. $k_3 \neq 0$ and $k_5 > 0$.

Since $\frac{dt_F}{ds} = 2k_2 t_N - k_1 t_F^2 + k_5 t_P$ & $\frac{dt_N}{ds} = k_1 t_F^2 - k_2 t_N - k_3 t_N t_P$ (eqs. 1.1&1.2), it follows

$$\frac{d(t_F + t_N)}{ds} = k_2 t_N + k_5 t_P - k_3 t_N t_P \qquad \text{(eq. 2.4)}$$

So, an equilibrium will occur between t_F and t_N if $k_2 t_N + k_5 t_P = k_3 t_N t_P$, such that $\frac{d(t_F + t_N)}{ds} = 0$.

Since $\frac{dt_N^*}{ds} = k_3 t_N t_P - k_4 t_N^*$ (eq. 1.3) & $\frac{dt_P}{ds} = 2k_4 t_N^* - k_3 t_N t_P - k_5 t_P$ (eq. 1.4), it follows

$$\frac{d(t_N^* + t_P)}{ds} = k_4 t_N^* - k_5 t_P. \qquad \text{(eq. 2.6)}$$

Thus, an equilibrium will exist if $k_4 t_N^* = k_5 t_P$, such that

$$\frac{d(t_N^* + t_P)}{ds} = 0. \qquad \text{(eq. 2.7)}$$

Given that $\frac{d(t_F + t_N + t_N^* + t_P)}{ds} = k_2 t_N + k_4 t_N^* - k_3 t_N t_P$ (eq. 1.5), and a steady state exists if $k_2 t_N + k_4 t_N^* = k_3 t_N t_P$, and

$$\frac{d(t_F + t_N + t_N^* + t_P)}{ds} = 0. \qquad \text{(eq. 2.8)}$$

Thus, in Scenario 3, an equilibrium can occur between t_F and t_N events (eq. 2.5) as well as between t_N^* and t_P events (eq. 2.7). Also, the overall system can exist in a steady state when $k_2 t_N + k_4 t_N^* = k_3 t_N t_P$ (see eq. 2.8).

When an equilibrium occurs between t_F and t_N events (eq. 2.5) as well as between t_N^* and t_P events (eq. 2.7), and when the overall system exists in a steady state whereby $k_2 t_N + k_4 t_N^* = k_3 t_N t_P$, then

$$\frac{d(t_F + t_N)}{ds} = k_2 t_N + k_5 t_P - k_3 t_N t_P \ (eq.2.4) \ = k_5 t_P - k_4 t_N^*. \qquad \text{(eq. 2.9)}$$

This equals (from eq. 2.6)

$$-\frac{d(t_N^* + t_P)}{ds} = -(k_4 t_N^* - k_5 t_P) \qquad \text{(eq. 3.0)}$$

Thus,

$$\left[\frac{d(t_F + t_N)}{ds} - \frac{d(t_N^* + t_P)}{ds} \right] = 0, \quad \text{when} \quad \frac{d(t_F + t_N + t_N^* + t_P)}{ds} = 0. \qquad \text{(eq. 3.1)}$$

196

If $\frac{d(t_N^* + t_P)}{ds} = 0$, then $k_4 t_N^* = k_5 t_P$, then based on eqs. 2.5 & 2.8,

$$k_4 t_N^* = k_5 t_P = k_3 t_N t_P - k_2 t_N \qquad \text{(eq. 3.2)}$$

Thus, $k_3 t_N t_P > k_2 t_N$ because if $k_3 t_N t_P = k_2 t_N$, then $k_4 t_N^* = k_5 t_P = 0$. Overall, the relative reaction rates at steady state are:

$$k_2 t_N = k_4 t_N^* = k_5 t_P = \frac{k_3 t_N t_P}{2} = \frac{k_1 t_F^2}{3} \qquad \text{(eq. 3.3)}$$

So the rate of $\underline{t_F + t_F \to t_N} > \underline{t_N + t_P \to t_N^*} > \underline{t_N \to t_F + t_F} = \underline{t_N^* \to t_P + t_P} = \underline{t_P \to t_F}$

Thus, when $k_1 = 1$, the relative rate constant values at steady state are:

$$k_2 = \frac{t_F^2}{3t_N}, k_3 = \frac{2t_F^2}{3t_N t_P}, k_4 = \frac{t_F^2}{3t_N^*}, k_5 = \frac{t_F^2}{3t_P} \qquad \text{(eq. 3.4)}$$

SUMMARY

In summary (Table 1), modeling (e.q. 3.1) shows that the overall system can exist in a steady state $\left[i.e., \frac{d(t_F + t_N + t_N^* + t_P)}{ds} = 0 \right]$. So, when the system is at steady state, two equilibria are present and $\frac{d(t_F + t_N)}{ds} = \frac{d(t_N^* + t_P)}{ds}$, and $\left[\frac{d(t_F + t_N)}{ds} - \frac{d(t_N^* + t_P)}{ds} \right] = 0$. Thus, analysis of the dynamics of the overall system indicates that two equilibrium reactions (between t_F & t_N and t_N^* & t_P) and the overall system can exist as a steady state.

Table 1. Summary of the Dynamics of Time Event Progression				
Scenario	Condition	Equilibrium between t_F and t_N events	Equilibrium between t_N^* and t_P events	Steady State
1	$k_3 = 0$, $k_5 > 0$	NO $\dfrac{d(t_F + t_N)}{ds} = k_2 t_N + k_5 t_P$	YES, if $k_4 t_N^* = k_5 t_P$	NO $\dfrac{d(t_F + t_N + t_N^* + t_P)}{ds} = k_2 t_N + k_4 t_N^*$
2	$k_3 > 0$, $k_5 = 0$	YES, if $k_2 t_N = k_3 t_N t_P$	NO $\dfrac{d(t_N^* + t_P)}{ds} = k_4 t_N^*$	YES, if $k_2 t_N + k_4 t_N^* = k_3 t_N t_P$
3	$k_3 > 0$, $k_5 > 0$	YES, if $k_2 t_N + k_5 t_P = k_3 t_N t_P$,	YES, if $k_4 t_N^* = k_5 t_P$	YES, if $k_2 t_N + k_4 t_N^* = k_3 t_N t_P$

Additionally, when the system is perturbed by setting k_3, or k_5, to equal zero, at least one of the equilibria will still persist. Indeed, an equilibrium between t_F and t_N (i.e. $\frac{d(t_F + t_N)}{ds} = 0$) is present when k_5, but not k_3, equals zero. Also, an equilibrium between t_N^* and t_P (i.e.

$\frac{d\left(t_N^* + t_P\right)}{ds} = 0$) is present when k_3, but not k_5, equals zero. Moreover, at steady state and when k_1 equals 1.0, expressions for both the relative reaction rates (eq. 3.3) as well as the relative rate constant values (eq. 3.4) can be derived for the system.

MODELING BASED ON A SPECIES REACTION GRAPH

Another way to understand the dynamics of the model is to plot it as a species reaction graph. The diagram shown in Figure 5 is a bipartite-type of graph with a distinct set of nodes representing the different time events (t_F, t_N, t_N^*, t_P) and the arrows connect them to the reactions they participate in. The graph maps out how the different time events interact within the reaction system [47].

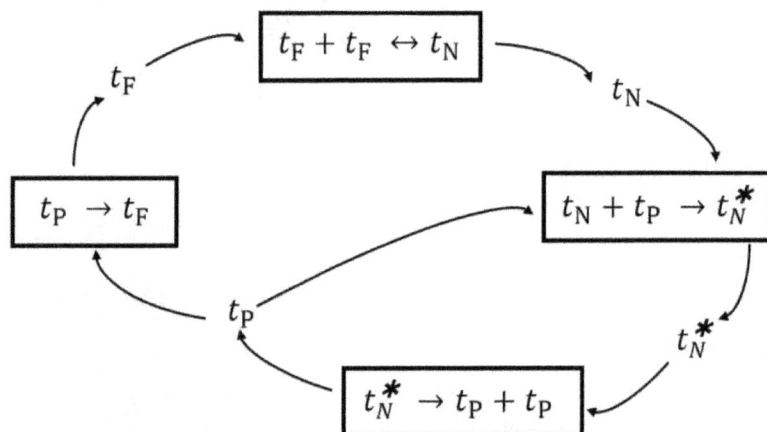

Figure 5. **A species reaction graph to visualize the relationships between the different time events involved in the progression of the reaction.** The Figure shows how each time event progresses to other events in the reaction process. Each node on the graph represents a different time event that participates in the kinetic reaction. The reaction includes $t_{\backslash mathrmF}$ events as reactants, t_N and t_N events as intermediates, as well as t p events as products. Arrows connect the nodes to the reactions they participate in, which depicts how the different time events interact within the reaction system. The arrows also illustrate the direction that the reaction progresses.

Analysis of this graph reveals several key points about the reaction pathways and dynamics.

1. The sequence of steps in the reaction system progresses asymmetrically and is cyclical.

2. Two reversible reactions ($t_F \leftrightarrow t_N$ & $t_N^* \leftrightarrow t_P$) exist involving two reversal reactions (comprised of a forward reaction & a reverse reaction).

3. An autocatalytic reaction occurs: $t_N + t_P \rightarrow t_N^* \rightarrow t_P + t_P$.

4. Two branch points exist: $t_F \leftarrow t_N \rightarrow t_N^*$ and $t_F \leftarrow t_P \rightarrow t_N^*$.

MODELING BASED ON A SIMPLE WALKING DIAGRAM

A further way to understand the model dynamics is by plotting the reaction system as a simple walking diagram (Figure 6). It illustrates the sequence of elementary steps whereby the reaction occurs. The plot also allows us to analyze the probabilities in the steps between time events of the system [48].

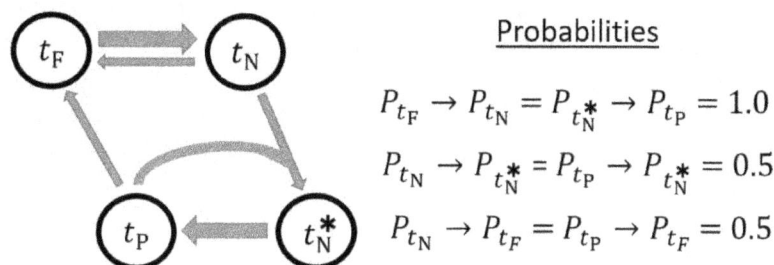

$$P_{t_F} \rightarrow P_{t_N} = P_{t_N^*} \rightarrow P_{t_P} = 1.0$$

$$P_{t_N} \rightarrow P_{t_N^*} = P_{t_P} \rightarrow P_{t_N^*} = 0.5$$

$$P_{t_N} \rightarrow P_{t_F} = P_{t_P} \rightarrow P_{t_F} = 0.5$$

Figure 6. **A simple walking diagram to illustrate *the* sequence of elementary steps by which the reaction occurs between the different time events.** Analysis of this graph reveals that three sets of equal probabilities exist when the system is a steady state.

The simple walking diagram allows the analysis of the two stable equilibrium states (between t_F & t_N events and between t_N^* & t_P events) that were identified in my above analysis of reaction kinetics (summarized in Table 1). In the equilibrium between t_F & t_N events, a time now (t_N) event is the only product generated from the reactants (t_F events) in the forward reaction so the probability is $P_{t_F} \rightarrow P_{t_N} = 1.0$. Since this reaction is a second order reaction $t_F + t_F \xrightarrow{k_1} t_N$ (rxn. 1.1), the *entropy should decrease* and the *potential energy should increase* in formation of a time now (t_N) event from interaction of the two time future (t_F) events.

For the stable equilibrium between t_N^* & t_P events, time past (t_P) events are the only products generated from the reactant (t_N^* event) in the forward reaction so the probability is $P_{t_N^*} \rightarrow P_{t_P} = 1.0$. This reaction involves an autocatalytic reaction whereby time past (t_P) events catalyze their own formation. In this reaction $t_N^* \rightarrow t_P + t_P$ (rxn 1.4), the *entropy should increase* and the *potential energy should decrease* in formation of two time past (t_P) events from a t_N^* event.

The switch between the two equilibrium states occurs at a branch point ($t_F \leftarrow t_N \rightarrow t_N^*$ and $t_F \leftarrow t_P \rightarrow t_N^*$) in both equilibria. The switch from one equilibrium state to the other equilibrium state occurs through two reactions: $t_N \rightarrow t_N^*$ and $t_P \rightarrow t_F$ (eqs. 1.3, 1.5). This mechanism allows the system to flip from one equilibrium state to the other equilibrium state. The simple walking diagram also shows the progression from one equilibrium to the other occurs in a cyclical asymmetric direction. My modeling results showing two stable equilibrium states indicates that the progression of time events occurs through a bistable dynamical system in which time events can transition from one equilibrium state to the other equilibrium state via hysteresis [49].

Thus, from a relational perspective, the occurrence of two stable equilibrium states shows that a bistable dynamic structure exists whereby the system exists in two different steady states under the same set of conditions. It also indicates that: 1) The system may switch between these different stable equilibrium states through the process of hysteresis, and 2) Oscillatory behavior can occur between the steady states in the system.

Discussion

The key finding from my study is that the happening of events in temporal relation to each other establishes bistable and self-sustaining dynamics of the Universe. Specifically, my study of the happening of events in temporal relation to each other and their respective points of reference reveals how the progression of events occurs at both the macroscale and microscale. Modeling from this relational perspective shows that the direction of progression of events is different at the macroscale and microscale because past events are immutable and irreversible. For the macroscale, the progression of events is from past to future because future events are caused by past events according to the temporal timeline.

For the microscale, the progression of events is from future to now to past because the progression is assumed to occur as a kinetic reaction and in kinetic systems a cause-and-effect occurs when a change in one event (the cause) triggers another event to occur (the effect). Thus, since past events are immutable, they are the effect, not the cause, of changes in other events. Indeed, in chemical reactions, events are causally linked through molecular collisions and the reactions occur due to collisions between reactant molecules that cause the irreversible formation of products. The direction of a chemical reaction in which products are irreversibly produced is the same as the direction of the arrow of time whereby past events are irreversible, fixed and unchangeable.

In my modeling, time now events occur as a transition step between time future events and time past events at the microscale. Specifically, time now events are happening at each and every point of reference in the Universe. In this way, the progression of events can be modeled based on transition state theory from chemistry whereby the progression of events is linked to interaction of future events, their energy of activation, catalysts that influence the reaction rate, the irreversible production of past events, and the increasing entropy of the Universe.

When time now events are modeled as happening at each point of reference and viewed from a relational perspective whereby events can only exist in relation to other events, modeling reveals several interesting dynamics that relate to fundamental concepts and paradigms in physics. The following aspects of model dynamics are discussed below:

1. Time past events at one point of reference progress to time future events at another point of reference.

2. The equilibrium between time future events and time now events.

3. Time past events are irreversibly produced at each point of reference.

4. Time past events autocatalyze their own production.

5. Bistability and hysteresis.

1. Time past events at one point of reference progress to time future events at another point of reference.

Notably, the direction of event progression in this reaction is the same as the asymmetric direction of progression of events (past to future) at the macroscale. In modeling, the reaction rate where time past (t_P) events progress time future (t_F) events is, according to rate law, a first order reaction controlled by the k_5 rate constant in the reaction $t_P \xrightarrow{k_5} t_F$ (rxn 1.5). The rate that time future (t_F) events are produced is described by the expression $\frac{dt_F}{ds} = 2k_2t_N - k_1t_F^2 + k_5t_P$ (eq. 1.1). Moreover, if future time events are continually produced from past time events, it could provide a mechanism that explains why the progression of time events is perpetual.

2. The equilibrium between time future events and time now events. In my model, an equilibrium exists between time future (t_F) events and time now (t_N) events. This is a reversible reaction between t_F events and t_N events, and is controlled by the k_1 and k_2 rate constants in the reactions $t_F + t_F \xrightarrow{k_1} t_N$ (rxn 1.1) and $t_N \xrightarrow{k_2} t_F + t_F$ (rxn 1.2). This equilibrium relates to several ideas in quantum mechanics including: i). Superposition, ii). Entanglement, iii). The wave function and its collapse, iv). Arrow of time, and v). Universe expansion.

Superposition. In my model, an equilibrium exists between time future (t_F) events and time now (t_N) events, such that an event will exist in both the future and present. Thus, in the equilibrium, time events exist in a superposition state whereby they are both t_F and t_N events. Specifically, in model scenarios 2 & 3, an equilibrium can exist between t_F and t_N events (see Table 1), and the model predicts that the system can reach a steady state. The concept of a superposition between t_F and t_N events might also relate to other superposition states in physics. For example, in quantum mechanics, superposition refers to a system that is able to be in multiple states at the same time [50-53]. The classic example is the concept of wave-particle duality whereby waves can exhibit particle-like properties while particles can exhibit wave-like properties which pertains to behavior of photons and electrons and other entities in the Universe. Perhaps in this equilibrium, time future events might have properties like a "wave" and time now events might have properties like a "particle" based on their differences in kinetic and potential energy (see Figure 4).

Entanglement. It can be envisioned that an equilibrium between t_F and t_N events might pervade throughout the Universe much like light and gravity operate on a universal scale [54]. In this view, t_N events at each point of reference are linked to all possible t_F events throughout the Universe. Furthermore, the interaction of two t_F events in a reaction that doesn't progress from t_N to t_P may dynamically be analogous to entanglement of two particles [55]. Thus, dynamics of two interacting t_F events could simulate two entangled particles. So, from the model, it might be predicted that two particles when entangled will exist in space throughout the Universe until they are measured.

If an interaction of two interacting t_F events dynamically simulates two entangled particles that are in an equilibrium with a t_N event and if the interaction of a t_N event with a t_P event portrays a measurement, then the model would simulate an "observation" or measurement of entangled particles.

The Wave Function and Its Collapse. The model dynamics simulate an equilibrium that is composed of a reversible reaction, i.e., a forward reaction and a backward reaction, which simultaneously are constantly fluxuating back and forth in a wave-like dynamic. This mechanism might relate to the wave function in quantum physics. The wave function is a mathematical description of a particle's quantum state as a function of momentum, time, position, and spin.

Moreover, in my model, the interaction of a t_N event with a t_P event that will irreversibly produce two t_P events might relate to the collapse of the wave function. The collapse of the wave function, refers to the process where a quantum system, initially in a superposition of states, transitions to a single, definite state when a measurement or observation happens. Like the interaction of interaction of a t_N event with a t_P event is modeled as the cause of t_P events, in quantum mechanics, the interaction between the quantum system and its environment is thought to be the cause of the collapse. Thus, modeling the interaction of a t_N event with a t_P event might provide a mechanism that helps explain the wave function collapse concept in quantum mechanics.

Arrow of time. The model can also help understand the asymmetric dynamics of time. If a superposition exists between t_F and t_N events that extends throughout the Universe, it would provide a ground state that is the same at every point of reference across the Universe. This state is a condition where time does not change it the superposition exists between t_F and t_N events, it only changes at the transition state when a t_N event progresses to a t_P event. It doesn't mean t_F and t_N are zero or that time is the same everywhere, it means that time doesn't change in the equilibrium between t_F and t_N events (i.e, $\frac{d(t_F+t_N)}{ds} = 0$). If this condition exists, it establishes a "when" from which time can progress asymmetrically at all points of reference in the Universe. This idea also pertains to time reversal symmetry which holds that the laws of physics can't distinguish between forward and backward directions in time. It sounds counter intuitive, but the "invariance" in symmetry might account for the "variance" in asymmetry. In this view, the ground state for progression of time events is based on a superposition involving a forward and backward direction in time. If so, then at each point of reference, the origin for asymmetric progression of time events might arise from this superposition.

One other prediction from time invariance in the equilibrium between t_F and t_N events that relates to quantum mechanics is my idea that there would be no change in time when the entangled particles are measured [44, 46]. That is, at the time of measurement, the particles are at exactly the same time as when they were entangled. If this is the case, there is no change in time of the particles upon measurement. So the argument that information travels faster than the speed of light doesn't make sense as speed is a function of change in time. Thus, the particles can be at different places in space but they are at the same "when" in time. Consequently, perhaps in quantum mechanics, we should be considering the temporal changes that do or don't occur when we are thinking about spatial non-locality.

Expansion of the Universe. My modeling results also indicate that the dynamics in Universe could exist as a network structure involving an equilibrium state between time future (t_F) events and time now (t_N) events. The existence of an equilibrium state could establish network structure whereby time events are dynamically interconnected between all points of reference throughout the Universe. In particle physics, the notion that the Universe consists of a network is not new. Indeed, several physicists have put forth ideas that spacetime can be viewed as a network in which the nodes are particles and these nodes are all connected by quantum entanglement [56]. Some theories even suggest that quantum entanglement might be responsible for the uniformity of the cosmos by connecting distant parts of the Universe [54-56]. In complex systems, the network of interactions can lead to growth of the system due to an increase in connections between nodes and hubs in the network [57]. Indeed, in model scenario 1 where an equilibrium cannot exist between t_F and t_N events (see Table 1), the model predicts that the system will expand. Thus, if the Universe is a dynamic network, it could provide an explanation for why the Universe is expanding.

3. Past time events are irreversibly produced at each point of reference. From this perspective, two past time events are immutable events irreversibly produced from a time now event at every point of reference, which provides a mechanism that might explain the increasing entropy of the Universe.

The conservation of energy and increase of entropy. The first law of thermodynamics states that energy cannot be created or destroyed, only transformed, while the second law of thermodynamics holds that in an isolated system entropy always increases [36]. In the model, time events are assumed to be parcels of energy which is akin to quantum mechanics where a quantum is a discrete quantity of energy rather than continuous ones [58-60]. In my modeling the progression of events, the energy fluxuates between kinetic energy and potential energy and overall energy is conserved. Also, in the model, the entropy decreases during the transition state in happening of time now events, but increases in the progression to time past events. This could provide a mechanism that might explain the increasing entropy of the Universe at the macroscale. In contrast, at the microscale, the entropy is assumed to decrease in time now events during the instantaneous transition state to time past events.

4. Time past events autocatalyze their own production.

Another question is how energy and matter might be conserved and be sustained in the Universe, rather than requiring a constant external input, which would be impossible assuming there is nothing outside the Universe?

Self-sustaining Steady State. The autocatalytic mechanism in the model might help understand how the Universe is self-sustaining. In my modeling, the autocatalysis by t_P events was incorporated as a mechanism to drive the continual progression of events. Findings from the model indicate that this autocatalytic mechanism accelerates the rate of the reaction by lowering the energy of activation in the transition state. However, this was from the microscale perspective of one point of reference and it doesn't necessarily indicate that the dynamics in whole Universe as a system are self-sustaining. So, a better question might be: does autocatalysis that increases the progression of events lead to a system that is self-sustaining? For this to happen, it would require that t_P events at all points of reference throughout the Universe function as autocatalysts. In that line of thinking, the Universe would need to be comprised of t_P events which are created by other t_P events in the Universe, such that as a whole, the entire system is able to catalyze its own production.

So, the autocatalyzed production of events at one reference point would need to promote the autocatalyzed production of events at another reference point, which would drive the perpetual progression of events in the system. The way that this could happen is if t_P

events at one point of reference are t_F events at another point of reference. Thus, the ability of t_P events to serve as t_F events at other points of reference and the ability of t_P events to autocatalyze their own production would theoretically make the whole system self-sustaining. In this way the Universe as a whole would dynamically be self-sustaining.

5. Bistability and hysteresis. My modeling results indicate that the Universe is self-sustaining and exists as a dynamic bistability that has two distinct, stable equilibrium states. In kinetic systems, bistability means that the system has two stable equilibrium states and can switch between the two states (termed hysteresis) [49]. In model Scenario 3 (see Table 1), the existence of a possible bistable state in the system was determined by showing that $\frac{d(t_F+t_N)}{ds} = \frac{d(t_N^*+t_P)}{ds} = 0$. Thus, the results showing two stable equilibrium states indicates that the progression of time events occurs through a bistable dynamical system in which time events can transition from one equilibrium state to the other equilibrium state via hysteresis. It also means that oscillatory behavior can occur between the steady states in the system. From a relational perspective, the dynamics of one equilibrium exists in relation to the other equilibrium and the two stable states are interdependent. In this context, it is important to understand the dynamics within each equilibrium as well as between the equilibrium states. The dynamics in one equilibrium involves a reversible reaction between t_F and t_N, and in the other equilibrium, it involves an autocatalytic reaction whereby t_P events catalyze their own production. The dynamics between the two equilibria also involves hysteresis that leads to a cyclical dynamic progressing in a one way direction.

Overall, the idea that the Universe exists as a dynamic relation between two stable equilibrium states and switches between them through hysteresis might provide a way to understand new mechanisms that explain the dynamic structure of the Universe. For example: 1) How the happening of events at different *points of reference* occurs in relation to each other might explain 2) How the happening of *frame of references* occurs in relation to other frames of reference. Addressing this gap-in-our-knowledge is important because many fundamental ideas and core concepts in quantum mechanics relate to frames of reference. The principle of Galilean invariance, states that the laws of mechanics remain the same in all inertial frames of reference. Galilean invariance is also one of the assumptions in Einstein's theory of special relativity along with the assumption that the speed of light is constant for all observers [61-63]. Noether's theorem can also be applied to Galilean symmetries based on a change in the frame of reference, leading to conserved quantities like momentum, energy, and angular momentum [64]. In this view, a frame of reference can move relative to other frames of reference and frames of reference moving at different speeds relative to each other is a key concept in physics, particularly in studies of motion and relativity. In relativity theory, Einstein links gravitation and acceleration of reference frames to rate of progression of time. For instance, time dilation refers to time passing more slowly in a moving frame than in a stationary frame, or in strong gravitational field than in a weak one. When considering energy of events in different reference frames relative to one another, it is predicted that a moving reference frame has more kinetic energy than a stationary reference frame, which according to the conservation of energy, would have more potential energy [65]. Based on my modeling, I previously conjectured that gravity and speed could increase the energy of activation and raise the height of the potential energy curve, which would delay the transition of events from t_F to t_P, and slow down progression of events [41, 46]. Based on my modeling herein on bistability, I surmise that a slow rate of progression would involve a decrease in the k_3 rate constant (Figure 4) and lead to a delay in transition between equilibrium states. So, when considering how interactions between two stable equilibrium states establish dynamics that are relative to one another, the dynamics of the Universe can be studied as the progression of events in terms of a kinetic reaction. Thus, the study of the dynamics of the Universe from a relational perspective could provide a way to understand new mechanisms and different aspects of the Universe, particularly in deciphering how its dynamics and physical phenomena are interconnected between the smallest and largest of scales.

References

1. Rovelli C. Helgoland: Making Sense of the Quantum Revolution (2022), Riverhead Books, pp. 1-256.

2. Borges JL. History of Eternity (1936), Viau y Zona, Buenos Aires.

3. La primera parte fue publicada en mayo de 1944 con el título "Una de las posibles metafísicas" en Sur, n ° 115, Buenos Aires, mayo de 1944, pp. 59-67. En 1946 publicó Nota preliminar y parte B de "Nueva refutación del tiempo". (la nota preliminar lleva la fecha 23 diciembre de 1946).

4. The Direction of Time. Edited by Sam Baron and Brigitte C. G. Everett, https://philpapers.org/browse/the-direction-of-time

5. Farr M. C-theories of time: On the adirectionality of time. Philosophy Compass 15: e12714, 2020, doi.org/10.1111/phc3.12714

6. Williams DC. The myth of passage. J Philos 48:457-472, 1951.

7. Smart JJC. Philosophy and Scientific Realism. New York, Routledge (1963).

8. https://iep.utm.edu/arrow-of-time/

9. Barbour J. The End of Time: The Next Revolution in Physics. (2001) Oxford University Press, pp 1-384.

10. https://en.wikipedia.org/wiki/Arrow_of_time

11. Hawking S. A Brief History of Time (1998), Bantam Publishing, pp. 1-212.

12. Rovelli C. The Order of Time (2018) Riverhead Books, New York, pp. 1-256.

13. Rovelli C, The Physics and Philosophy of Time, The Royal Institute, July 13, 2018.

14. Carroll SM, Chen J. Spontaneous Inflation and the Origin of the Arrow of Time, Archived November 18, 2017 at the Wayback Machine

15. Carroll S. Something Deeply Hidden: Quantum Worlds and the Emergence of Space-time (2020), Dutton Publishing, pp. 1-368.

16. Penrose R. Cycles of Time: An Extraordinary New View of the Universe (2012), Vintage Publ., pp. 1-304.

17. The Arrow of Time, The Internet Encyclopedia of Philosophy (IEP) (ISSN 2161-0002), https://iep.utm.edu/arrow-of-time/

18. Smolin L. Time Reborn: From the Crisis in Physics to the Future of the Universe (2013), Mariner Books, pp. 1-357.

19. Ben-Naim A. Entropy and Time. Entropy (Basel). 22:430, 2020. doi: 10.3390/e22040430.

20. Ben-Naim A. Time's Arrow (?): The Timeless Nature of Entropy and the Second Law of Thermodynamics. (2018) Lulu Publishing Services, pp 1-218.

21. https://www.britannica.com/dictionary/how definition

22. https://www.merriam-webster.com/dictionary/time

23. https://www.merriam-webster.com/dictionary/event

24. Zwart PJ. The flow of time. Synthese 24:133-158, 1972.

25. https://www.historyskills.com/historical-knowledge/causes-and-consequences/?srsltid=AfmBOorLZhhav-IOm6wavfraO47K6pETGhDeVhWaQtTU-1iy6ADhzp-o

26. Molet M, Miller RR. Timing: an attribute of associative learning. Behav Processes 101:4-14, 2014. doi: 10.1016/j.beproc.2013.05.015.

27. Space and Time: Minkowski's Papers on Relativity, published by the Minkowski Institute. The image had appeared on the cover of The Mathematical Intelligencer, Volume 31, Number 2 (2009).

28. Feynman R. (1949). "The Theory of Positrons". Physical Review. 76 (6): 749-759. Bibcode:1949PhRv...76..749F. doi:10.1103/PhysRev.76.749. S2CID 120117564.

29. Hermann, G. (2016). Natural-Philosophical Foundations of Quantum Mechanics. In: Crull, E., Bacciagaluppi, G. (eds) Grete Hermann - Between Physics and Philosophy. Studies in History and Philosophy of Science, vol 42. Springer, Dordrecht. https://doi.org/10.1007/978-94-024-0970-3_15.

30. Angulo D, Thompson K, Nixon VM, Jiao A, Wiseman HM, Steinberg AM. (2024) Experimental evidence that a photon can spend a negative amount of time in an atom cloud. https://arxiv.org/abs/2409.03680

31. Fullwood J, Vedral V. (2025) Geometry from quantum temporal correlations. arXiv preprint arXiv:2502.13293.

32. Sagan C. (1994). Pale Blue Dot: A Vision of the Human Future in Space (1st ed.). New York: Random House. ISBN 0-679-43841-6.

33. Beynon JH, Gilbert JR. Application of Transition State Theory to Unimolecular Reactions: An Introduction. John Wiley & Sons, New York. 1984.

34. Glasstone S, Laidler KJ, Eyring H. The Theory of Rate Processes: The Kinetics of Chemical Reactions, Viscosity, Diffusion and Electrochemical Phenomena. McGraw-Hill Book Company Inc, New York & London. 1941.

35. Moore JW, Pearson RG. Kinetics and Mechanism. John Wiley & Sons Inc, New York, NY, USA. 1981.

36. Sheehan WF. Physical Chemistry 2nd edition. Allyn & Bacon Inc, Boston. 1970.

37. Eyring H. The Activated Complex in Chemical Reactions. J Chem Phys 3:107-115. 1934.

38. Evans MG, Polanyi M. Some applications of the transition state method to the calculation of reaction velocities, especially in solution. Trans Faraday Soc 31:875-94. 1935.

39. Laidler K, King C. Theories of chemical reaction rates. R. E. Krieger Publishing Co., New York. 1979.

40. Laidler KJ. Just what is a transition state? J Chem Educ 65:540-2. 1988.

41. Laidler KJ. Theories of chemical reaction rates. R. E. Krieger Publishing Co., New York. 1979.

42. Boman BM. A Simple Mathematical Model for the Asymmetry of Time. Presented at the Fifth International Conference on the Nature and Ontology of Spacetime, Varna, Bulgaria, May 16, 2018.

43. Boman BM. A Simple Mathematical Model for the Asymmetry of Time. arXiv:2102.08826v1, Feb 14, 2021.

44. Boman BM. Modeling spacetime based on transition state theory. In: Slagter RJ, Keresztes Z (eds) Spacetime 1909-2019. Minkowski Institute Press, Montreal. 2020. ISBN 978-1-927763-54-4, pp179-206.

45. Boman BM. Becoming as a Transition State in Spacetime. Presented at the Sixth International Conference on the Nature and Ontology of Spacetime, Varna, Bulgaria, September 12, 2022.

46. Boman BM (2024) Modeling the Progression of Time as an Autocatalytic Reaction In: Ling E, Piubello A (Eds), Spacetime 1908-2023, Minkowski Institute Press (July 9, 2024) p204-226.

47. Shinar G, Feinberg M. Concordant chemical reaction networks and the Species-Reaction Graph. Math Biosci. 2013 Jan;241(1):1-23. doi: 10.1016/j.mbs.2012.08.002.

48. Bender EA.; Williamson SG. (2010). Lists, Decisions and Graphs with an Introduction to Probability https://cseweb.ucsd.edu/~gill/BWLectSite/Resources/LDGbookCOV.pdf.

49. https://en.wikipedia.org/wiki/Bistability

50. Messiah A. (1976). Quantum mechanics (2 ed.). Amsterdam: North-Holland. ISBN 978-0-471-59766-7.

51. Dirac PAM. (1947). The Principles of Quantum Mechanics (2nd ed.). Clarendon Press.

52. Monroe, C.; Meekhof, D. M.; King, B. E.; Wineland, D. J. (24 May 1996). "A "Schrödinger Cat" Superposition State of an Atom". Science. 272 (5265): 1131-1136. doi:10.1126/science.272.5265.1131.

53. Streltsov A, Adesso G, Plenio MB. (2017). "Colloquium: Quantum coherence as a resource". Reviews of Modern Physics. 89 (4): 041003. arXiv:1609.02439. Bibcode:2017RvMP...89d1003S. doi:10.1103/RevModPhys.89.041003. ISSN 0034-6861. S2CID 62899253

54. Feynman R. (1970). The Feynman Lectures on Physics. Vol. I. Addison Wesley Longman. ISBN 978-0-201-02115-8.

55. Horodecki R, Horodecki P, Horodecki M, Horodecki K. (2009). "Quantum entanglement". Reviews of Modern Physics. 81 (2): 865-942. doi:10.1103/RevModPhys.81.865.

56. Halperin, A. The Network of Time: Understanding Time & Reality through Philosophy, History and Physics (2020) ISBN: 9798672645117

57. https://science.nasa.gov/what-is-the-spooky-science-of-quantum-entanglement/

58. Maldacena J. Entanglement and the Geometry of Spacetime, (2103, https://www.ias.edu/about/publications/ias-letter/articles/2013-fall/maldacena-entanglement).

59. https://en.wikipedia.org/wiki/Quantum_entanglement

60. Barabasi AL, Albert R. Emergence of scaling in random networks. Science. 1999 Oct 15;286(5439):509-12. doi: 10.1126/science.286.5439.509.

61. Einstein A. (2015) Relativity: The Special and the General Theory, 100th Anniversary Edition (Princeton University Press, Princeton).

62. Davies P. (1995) About Time: Einstein's Unfinished Revolution, (Simon & Schuster, New York).

63. Muller RA. (2016) Now: The Physics of Time, (WW Norton & Company, New York & London).

64. Noether, E. (1918). "Invariante Variationsprobleme". Nachrichten von der Gesellschaft der Wissenschaften zu Göttingen. Mathematisch-Physikalische Klasse. 1918: 235-257.

65. Dourmashkin P. Classical Mechanics, https://isaacphysics.org/concepts/cp_frame_reference, Physics LibreTexts.

10   Dynamical Structured Spacetime Realism (DSSR): The Emergence of Low-Dimensional CPT-Symmetric Spacetime

Charlie Dawson

Abstract   In this paper, we introduce a novel ontological model called **Dynamical Structured Spacetime Realism (DSSR)**. This theory aims to bridge the apparent discrepancy between the higher-dimensional wavefunction of quantum theory and the 3D macroscopic reality we observe. DSSR proposes that 3-dimensional space emerges from *knotted phase singularities* in an *n*-dimensional excitable vacuum, forcing all matter to be localized on a 3D "brane." Furthermore, time emerges via an *entropy exchange* mechanism, in which local fermion-generating excitations can remain in low-entropy states while the global environment's entropy increases. This approach clarifies how wavefunction realism, structural realism, and spacetime state realism might co-exist, without invoking extra dimensional compactification. Crucially, DSSR preserves the second law of thermodynamics by distinguishing *local* vs. *global* entropy changes, thus avoiding contradictions. Finally, I discuss open questions: (*i*) the uniqueness of our particular 3D universe and (*ii*) whether multiple 3D embeddings might coexist in a higher-dimensional bulk. I also provide possible future directions, such as formalizing these ideas with topological quantum field theories and exploring experimental/observational signatures of knotted vacuum excitations.

1   Introduction

We continually struggle to reconcile the *probabilistic* nature of the wavefunction in quantum mechanics with the seemingly *deterministic* laws of the macroscopic world. Scientific realists debate to what extent the wavefunction should be taken as ontologically fundamental. Some interpretations, such as *Wavefunction Realism (WFR)*, treat the wavefunction as an entity existing in a high-dimensional configuration space. Others, such as *Ontic Structural Realism (OSR)*, posit that the *structures* described by mathematics are what truly exist, transcending familiar 3D objects.

The challenge that all realist theories face is to go beyond mathematical formalizations and to provide a means by which the world we experience with static macro-objects and deterministic cause and effect mechanics co-exists with the probabilistic world of quantum mechanics. This is a challenge that has not reached a definitive conclusion.

The question remains: *Why do we experience only three spatial dimensions when some theories imply many more?* String-theory approaches have suggested *compactified* extra dimensions, but that does not explain why only three are "large."

Kyley Ewing (Ed.), *Spacetime Conference 2024. Selected peer-reviewed papers presented at the Seventh International Conference on the Nature and Ontology of Spacetime, 16 - 19 September 2024, Albena, Bulgaria* (Minkowski Institute Press, Montreal 2025). ISBN 978-1-998902-44-6 (softcover), ISBN 978-1-998902-45-3 (ebook).

DSSR: A Topological Mechanism. *Dynamical Structured Spacetime Realism (DSSR)* offers an alternative: 3D space arises from *knotted scroll-wave filaments* that confine fermions (the matter particles) to exactly three dimensions. Knots, in the strict sense, are stable only in 3D, so matter cannot exist in higher or fewer spatial dimensions. By extension, all macroscopic objects composed of matter must inhabit a 3D membrane in an n-dimensional bulk. DSSR further proposes that *time* emerges from an *entropy exchange* process: fermion-generating excitations remain locally in lower-entropy states, but the rest of the universe increases in entropy, preserving the second law.

Emergent Spacetime in DSSR. Quantum mechanics (QM) is rife with observations and measurements that can't be reconciled with our common experience. This has sparked philosophical debate regarding if and to what degree observed quantum phenomena should be taken as representative of objective reality. A common theme that is present in all interpretations of QM is the central role of the wave function. It's important to take note of 2 different ways that the wavefunction is commonly conceptualized, (1) the wavefunction as a probability amplitude in configuration space, (2) the wavefunction as physically real excitations of quantum fields. Only the former is an accurate definition of a wavefunction. The latter is often, and erroneously, included in conceptualizations of the *universal* wave function.

The universal wave function isn't a *wave in a field*. It is a solution to a partial differential equation that looks like a wave when plotted on a coordinate graph. The waves, or excitations, in QFT are regions of the vacuum energy that are in a state of oscillation which dictates a localized spike in the energy of the field at that location.

The ambiguity is understandable considering that the wave on a coordinate graph is associated with spacetime coordinates where the real particle is likely to be measured. While the coordinates of the wave function can be associated with real spacetime coordinates, it's a mistake to say that the probability amplitudes of the universal wave function are the undulations of the quantum fields. The two are very closely related. They are perhaps codependent. Codependent does not mean synonymous, and the distinction is important.

The wavefunction's central role in QM is the foundation for Wave Function Realism (WFR). WFR is not a single interpretation of quantum theory. It is an adaptable framework that posits the ontological underpinnings of the observable universe are founded in a quantum wave function that exists in a higher dimensional space.[18] QM describes particles in states of superposition. Because we do not experience a macroworld that exists in a superposition, the wave function is said to have collapsed, bringing particles into a classical state that can be assigned definitive spatiotemporal coordinates. The wavefunction requires additional degrees of freedom to reconcile the pre-collapse superposition states with the classical states of 3+1D spacetime.

In contrast to WFR is Ontic Structural Realism (OSR), which suggests structure is the foundation of reality. OSR posits that as our understanding of the natural world advances, the parts of our knowledge that are retained are mathematical and observational knowledge that describe structures, leading to the conclusion that structure is all there is to know about the ontological nature of reality.[1]

The challenge for these realist theories and others like them is to account for the juxtaposition of our observed low-dimensional reality and the apparent high-dimensional behavior of the wavefunction. The approach that many realists have taken is first attempting to define a rigorous mathematical space that the wavefunction exists in (a wave in configuration space, a ray in Hilbert space, etc.). Without a solid conceptual understanding that tethers these mathematical abstractions to the immediately apparent world, the result of this approach has been vague, loose ideas that lack a plausible explanation of how these worlds cohabitate.

Some string theorists have proposed compactified dimensions to explain the dimensions which are apparently missing from the world of our common experience.[21] Despite impressive mathematical rigor, string theories that rely on compactified dimensions beg questions as to how or why some dimensions become compactified while others are present in the observable macro-world. They offer only explanations of what a compactified dimension might be, not a behavioral or mechanistic explanation of compactification.[24]

Ontological models that make up the contemporary conversation each have something that they are asserting to be the foundation of reality. Whether it's the wave function, structures, states of spacetime, etc., there is something being asserted as the most fundamental element of reality. However, each of these proposed fundamental components are themselves an abstraction. This isn't necessarily problematic if these abstractions truly reflect what is ontologically fundamental.

DSSR is a theory that embraces the higher dimensionality of contemporary theories. Its aim is to make sense of both the n-dimensional aspect of reality that has been revealed to us through observations of the quantum world and the aspect of reality apparent through immediate observations of the low-dimensional world that we experience every day. DSSR first considers what things like the wave function and structures are an abstraction of and then posits that an abstraction can serve as the mechanism that relates the higher-dimensional and lower-dimensional versions of reality to one another. DSSR uses the mathematical abstractions that describe our universe for what they are: *logical constructs that can provide insights about reality not immediately apparent from our vantage point as observers embedded in the system being observed.* By doing this we are able to distill the useful aspects of these abstractions and use them to build a mechanism that can make sense of equally real observations of the universe, whether at the quantum or macro scales.

The novelty of DSSR is not in the particular abstraction that is posited[1], but rather in (1) the suggestion that this abstraction will necessarily relate the nD bulk to the 3D world, (2) in proposing an exchange of entropy in physical systems, from the fundamental particles that systems are made of, to the physical systems that are apparent in the macro world, and that the temporal dimension is an emergent property of this entropy exchange mechanism.

In this paper we introduce a novel model of an emergent spacetime. This new model will provide a plausible mechanism that can get us from higher-dimensional spaces to the familiar 3 spatial dimensions. We avoid discussions of configuration space, Hilbert space, etc. This may seem like a strange approach at first, as much of the contemporary discussion of ontology seems to be centered around which type of mathematical space should be used to formalize the wave function. The upshot will be that in remaining untethered from any

---

[1]This is to say that we did not develop the field of scroll waves, nor did we generalize scroll waves as local solutions to field equations. We applied the work that has been done in the field of scroll wave dynamics to posit these as solutions to fermionic fields and developed the implications of this application.

specific mathematical form we will be able to build a new framework from the ground up based on a conceptual understanding that is informed by rigorous, abstract math. This model will make intuitive sense of observations of the quantum world that seem to require further explanation, such as superposition, and our observations of the world we experience every day.

We will begin by describing how n-dimensional wave dynamics create 3-dimensional phase singularities. This is the mechanism to get from the higher-dimensional wave function to low-dimensional space. In other words, we will start by describing how the "space" part of spacetime emerges from the higher-dimensional wavefunction. This leaves the temporal component of spacetime still to be accounted for.

Surprisingly, we will see that time is an emergent property of space, as opposed to the more common notion of the opposite. We show that the 4-dimensional extension of these knotted phase singularities which have become embedded in 3 dimensions dictates that these excitations, when viewed as localized systems, evolve to states of decreasing entropy. Entropy is then seen to be an exchange between the field excitations that generate fermions, and the systems that exist in the universe which are fundamentally composed of fermions. We show that space and time are emergent properties of wave dynamics in a higher-dimensional bulk. From high-dimensional wave dynamics, knotted phase singularities emerge that are inherently 3-dimensional. From these dynamical structures that are embedded in 3 dimensions, the temporal dimension emerges as an exchange of entropy. Spacetime emerges from an n-dimensional bulk as a discrete symmetry, CPT symmetry, recognized by embedded observers.

The paper proceeds as follows:

§2 gives background on quantum fields, fermions, spinors, and excitable media, clarifying the main concepts.

§3 describes how *knotted scroll waves* form in 3D excitable media, linking them to spin-1/2 fermions.

§4 analyzes *local vs. global* entropy changes, ensuring no violation of thermodynamics.

§5 discusses how *time* emerges through entropy flow rather than violating known physics.

§6 addresses open questions, including "Why our 3D?" and how multiple 3D slices might (or might not) align into one observed continuum.

§7 concludes, summarizing the model's contributions and suggesting future research directions.

2 Essential Physics and Conceptual Tools

2.1 Quantum Field Theory and Fermions

The *Standard Model* describes fundamental particles as excitations of underlying *quantum fields*. Fermions (e.g. electrons, quarks) have half-integer spin and obey the Pauli exclusion principle. They are often represented by *spinor* solutions of the Dirac equation. Bosons, meanwhile, are integer-spin force carriers (e.g., photons). In standard usage:

- **Vacuum Energy.** The vacuum is often seen as the *lowest-energy state* of the quantum fields. However, in DSSR, we treat this vacuum as a *dynamic, excitable medium*, with the baseline energy capable of forming local wave-like excitations.

- **Spinors.** Unlike vectors, spinors require a 720° rotation to return to the same state. After 360°, the spinor is effectively multiplied by −1. This *duality* is crucial to understanding fermion behavior.

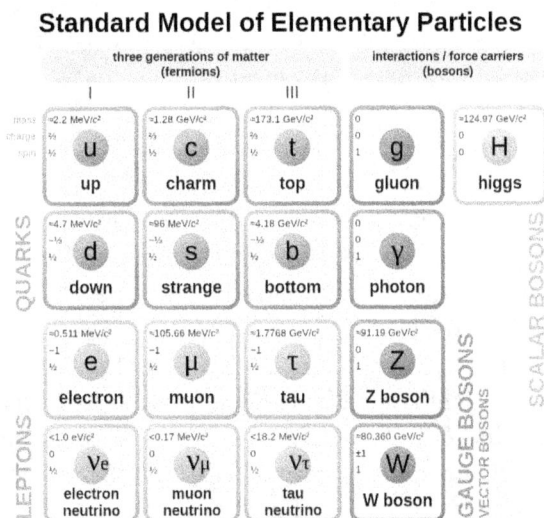

Figure 1: The Standard Model [4]

Quantum Field Theory (QFT) and The Standard Model of Particle Physics.
The standard model is like a periodic table for the fundamental particles of physical reality. (Fig. 1) It is divided into 2 main categories: fermions and bosons. Fermions are fundamental particles of matter. All the 'stuff' in the universe is made of fermions. If it is made of atoms, then it's fundamentally made of fermions. Bosons are force carrier particles. They are quantized bits of pure energy. They facilitate interactions between fermions by transmitting forces between fermions. Fermions, and specifically the structure of excitations that generate fermions, will be the focus of this paper.

QFT is a theory that posits that the universe is filled with fields and that particles are oscillations in these fields. It is a field theory. The standard model describes a field that is associated with every fundamental particle. For example, in the standard model there is an electron field, and every electron that exists, has existed, or will exist in the universe is an oscillation in this singular field.[22]

More accurately, particles that are generated by fields in QFT aren't just oscillations. They are excitations of the field. Particles are regions of the field where there is a localized energy spike compared to the baseline energy value associated with that field. (Fig. 2)

215

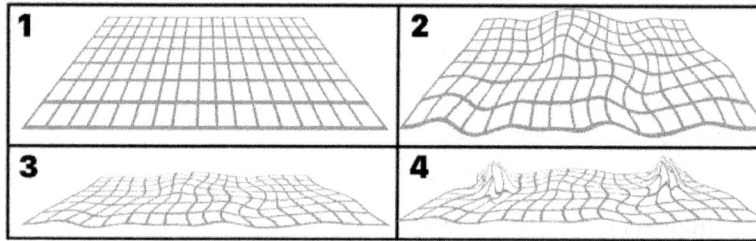

Figure 2: In QFT particles are localized excitations of fields

The *fields* of QFT are a mathematical abstraction. The particles that are generated from these fields are tactile. These particles interact with light. They interact with one another. They are as real as anything that exists in the material universe. Electrons are generated as oscillations in the electron field, but there is no electron field that is *materially real* in the same sense that an electron is *materially real*. There is no electron field where, in the absence of an oscillation that generates an electron, the field can interact with light. The electron field has the potential to bring an electron into the materially real universe. Quantum fields have the potential to generate materially real particles, but in the absence of those particles, fields are invisible and immaterial. The question we should ask is, "What are the quantum fields an abstraction of?"

The answer is that the fields of QFT are an abstraction of the *vacuum energy*. It's important to take note of the difference in the way that DSSR treats vacuum energy and the vacuum state of a field in QFT. In QFT the vacuum state is the state of the field in its lowest energy state. Throughout this paper we will discuss the vacuum energy as the baseline amount of undulating, dynamic, background energy that exists throughout the universe.[2]

The most apparent analogy is to compare the vacuum energy to the water in a fish tank. Fish might swim around in a tank full of water, largely unaware that they are in water. From our perspective outside of the tank it is clear to us that the tank is filled with water. It is apparent because we are looking at the tank from a place that is not filled with water. The difference is that a fish isn't made from the water in the tank. The decorations and glass and filter are not the water in a state of oscillation. Still, there is something to be taken from this imperfect analogy, which is that we are observers that are embedded in and made of vacuum energy, in a universe that is filled edge to edge with vacuum energy.

There is certainly a case to be made that the material ontology of everything is energy. Bosons are pure energy. Fermions are matter. Matter has intrinsic mass. There is a mass-energy equivalence described by $E = Mc^2$. If matter is annihilated it will release the energy that is stored as potential energy in the form of intrinsic mass and be converted to boson radiation, or pure energy.[10] This isn't a very interesting assertion, however, as it does nothing to describe our apparent reality in any sort of ontologically significant way that might help us understand how we get the material universe from vacuum energy.

Typically, when we think of the most fundamental components of physical reality we think of the particles of the standard model. Without the particles of the standard model,

---

[2]DSSR treats vacuum energy something like the energy density of the false Higgs vacuum. This is a metastable vacuum posited in theories of false vacuum decay.[3]

there isn't anything that we could assert to be physical reality. The particles of the standard model are generated as oscillations in quantum fields associated with each particle. These fields are an abstraction of the vacuum energy. So, when we focus solely on the vacuum energy as being ontologically fundamental, we aren't able to posit anything of real interest pertaining to physical reality. The vacuum energy doesn't interact with light, so we can't detect it. We are made of vacuum energy in a state of oscillation, in the form of fundamental particles, so we cannot observe it from a perspective outside the vacuum energy to assert anything that might only be apparent from that perspective.

We should consider what is fundamentally real to be something dynamic, something that is rooted in the fields, which generates the fundamental particles. What we're looking for is a dynamic process. But, any kind of process, or anything that exists over time, cannot get us from an n-dimensional bulk to our 3-dimensional reality. We live in a world with 3-spatial dimensions, so any kind of dynamical process requires an additional degree of freedom to be dynamic instead of static.[3] What would solve our problem is a dynamic process that somehow plays out specifically in 3-dimensional spaces. Because the n-dimensional bulk might have any arbitrary number of dimensions, we would need to be tied to 3 dimensions of space in an unavoidable, mathematically rigorous means that could be known a priori.

The kinds of "structures" that structural realists suggest are retained as understanding advances aren't "structures" in the everyday sense. Early versions of structural realism proposed epistemic structures. Contemporary ideas, such as OSR, advocate for retaining the essence of relationships described in abstract mathematics, preservation of symmetry groups (e.g. any future *Grand Unified Theory (GUT)* will need to preserve existing $(U(2) \times SU(2) \times SU(3))$ gauge symmetries), the realness of individual objects or relationships, whether relationships necessitate relata, etc.[16]

In contrast, the 'structure' of DSSR refers to geometric/topological structure. We are looking for something that can come from a higher dimensional space and through a dynamical process gets confined to 3 spatial dimensions. Then, within the context of its confinement to 3 spatial dimensions, allows for the continuation of the dynamical process that causes confinement to 3 dimensions, thus allowing extension back into a 4th dimension. We need a physical structure that is inherently 3-dimensional.

2.2 Spinor Duality and Advanced Introduction

Because spinors transform in ways less familiar than vectors, it helps to explain them *before* referencing their dual nature in the text. A spin-$\frac{1}{2}$ particle (e.g. an electron) is described by a four-component Dirac spinor that effectively contains both the particle and antiparticle degrees of freedom, each with two spin states. The *geometric* manifestation of this is the well-known "$\pm 2\pi$" rotation nuance. Later, in §3, we connect this half-integer rotation to *topological filaments* in 3D.

Solutions to field equations represent fermions as spinors.[7] Spinors are an incredibly useful tool in describing the quantum world, but they are just that, a tool. We seek a conceptual understanding of the particles that make up all matter in the universe. Spinors are not

---

[3]At the moment we are concerned with why the universe has 3 spatial dimensions.

fermions, but there is a reason that math dictates these counterintuitive mathematical objects to be the representation of fermions. Pinning down what it is about spinors that makes them a useful description of fermions can help us develop the conceptual understanding we are after.

We have established that quantum fields are mathematical abstractions of the vacuum energy. It stands to reason that if equations describe particles generated from quantum fields, the solutions to those equations would also be mathematical abstractions. Our goal is to progress our understanding of the universe. With that end in mind, we shouldn't be satisfied with quantifying the world in the language of abstract math. We should look deeper. The universe reveals itself to us through math, and we must interpret that math to understand the nature of the universe. Spinors are very strange objects that have a kind of intrinsic duality to them. The nature of a spinor as an object is unlike the common nature of objects in the material universe. There is a kind of dynamic duality, a sort of Janus perspective, uncoupled from the local environment, that keeps spinors relegated to the world of logic. This internal, self-symmetry cannot be reconstructed in the macroscopic, material world.

Likewise, quantum particles behave in ways that seem to indicate a duality to the material universe that deserves an explanation that we have been unable to satisfactorily provide. It is our job to look at spinors and figure out what it is about the universe that is being told to us through the inclusion of spinors in its description by mathematical quantification. Fermions and spinors are considered synonymous in the world of mathematical physics. The generation and annihilation operators in the Dirac equation might yield some positive result of electrons that are generated from the electron field, for instance. The equation will give this result as the generation of *spinors* in the electron field. This is the case for all fundamental particles of matter, and so fermions are often referred to as *spinoral matter*.

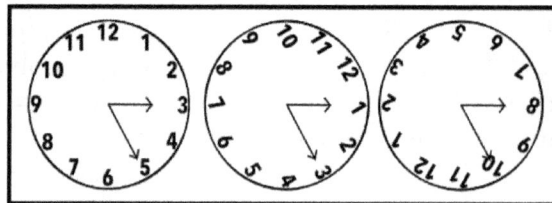

Figure 3: The hands of the clock remain fixed, and the clock face rotates behind them

Spinors live in vector spaces over the complex numbers. Vectors and tensors also inhabit these spaces. The distinction between spinors and their vector space kin becomes apparent when coordinate systems are rotated. If we rotate the coordinate system, vectors will be apathetic to the rotation. They will point to different coordinates, material stress will be described in a different coordinate direction, but the local geometry will be unaffected. We can think of this like moving the face of a clock behind the arms. In Figure 3 the times on the clocks read 3:25, 1:15 and 8:20, but the arms of the clock show the hour hand pointing to the right and the minute hand rotated 60° clockwise from the hour hand. Only the coordinates of the clock face have changed.

Spinors aren't only sensitive to the ending coordinate rotation the way that a vector is.

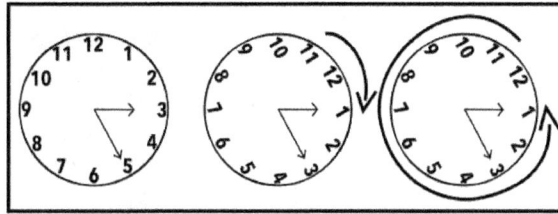

Figure 4: There are two ways that the clock face can be rotated to achieve an equivalent ending orientation

Spinors are sensitive to the way the rotation occurred. With a vector we can say that the clock used to read 3:25 and now reads 1:15, and that's all that matters. A spinor accounts for the way the clock face moved to reorient the coordinate system. Specifically, there are two ways that the clock face could have been rotated to result in the clock reading 1:15 (Fig 4), and a spinor needs to account for both possibilities. The duality of a spinor is that it contains two opposite coordinate rotations at the same time.

Although not a rigorously accurate definition, it's not uncommon to talk about spinors as being the square roots of vectors. It's useful to think of them as square roots in the sense that in complex number systems there is some ambiguity when we look for a square root. For example, since both $(-2) \times (-2) = 4$ and $2 \times 2 = 4$, there is some ambiguity about the problem $x = \sqrt{4}$ (or $x^2 = 4$). This is a useful way to think of spinors because in some sense spinors are vectors that contain their own negative, similar to the way that either 2 or -2 equivalently satisfy the equation for x.

There is a characteristic duality of an object that contains a reflection of itself, representing itself as both positive and negative simultaneously, inherent to a spinor. We suggest this key characteristic of spinors should be conceptualized to understand the nature of a fermion. Fermions have a dynamic symmetry. They are ever-changing reflections of themselves. This kind of symmetry requires both the positive and negative versions of something are simultaneously embodied in the same unique coordinate. No more than one material body can occupy a single spatiotemporal coordinate, so conceptualizing this kind of symmetry and applying it to our understanding of fermions can be difficult.

To do this we need to consider how the spinor is reoriented by continued rotation. If we imagine standing in front of a mirror, if we turn around and stop after 360° of rotation, we will be reoriented facing the mirror again. A spinor needs to be rotated 720° before it returns to its original orientation. After 360° a spinor is oriented as the negative of its original orientation. It might be helpful to think of a spinor as being turned inside out as it rotates, so that after a single rotation the spinor is facing the same direction, but inside out. As it rotates all the way around again, it ends up facing the same direction but now it's right side out. This isn't a rigorous description, but it can be a useful way to think of this sort of superimposed symmetry.

Conceptualizing Spinor Duality via Mobius Strip analogy. Something that is often used to convey the strangeness of a spinor is a Mobius strip. In Figure 5 we see a plane with blue arrows on top and red arrows underneath. Frame 2 shows that both sets of arrows are facing the same edge of the strip. In frame 3 the strip is twisted so that one end is flipped

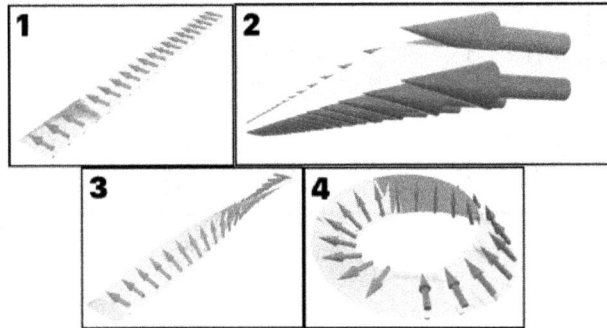

Figure 5: Mobius Strip

over. Notice that now on one end of the strip the blue arrows are still on top, and on the other end the red arrows are now on top. In frame 4 the strip is curled around on itself so that both ends connect to one another. If you imagine starting in front of the first blue arrow, tracing out the edge and continuing around, after 360° around the strip, you will find yourself on the same edge of the strip, but now behind the blue arrow. After following the edge of the Mobius strip around for another 360° in the same direction, you'll find yourself on the same edge, but now back in front of the blue arrow, at the same place you started. Just like a spinor, you are reoriented in your original configuration after 720° of rotation.

A spinor is only a mathematical construct. However, this invention of logic has a strange characteristic, which is that they are a reflection of themselves, both the positive and negative, contained in the same space. It is this strange characteristic that makes them a useful stand-in for understanding fermions. Fermions are more than just a logical construct. Fermions are real. There is a characteristic of a fermion that is familiar as the idea of something and its reflection being contained in the same space. This should be our takeaway from spinors. There is a characteristic of fermions revealed through mathematical constructs that have a characteristic duality foreign to our common experience of the world.

2.3   Excitable Media and Localized Oscillations

An *excitable medium* is one in which oscillations actively change the state of the medium itself. The Belousov–Zhabotinsky (BZ) chemical reaction, for instance, visibly oscillates between color states, each stable for brief periods. In 2D, we see *spiral waves* around a point-like core. In 3D, those waves become *scroll waves*, with a 1D "filament" serving as the phase singularity.

Contrasting Oscillations with Excitations.   Excitations are oscillations in an excitable medium. For an oscillation to be an excitation, the media housing the oscillation must have the capacity to be in a state of excitation. The medium in which the excitation is propagating must have more than one stable state that can be sustained. As the system evolves towards equilibrium the media will periodically oscillate between different stable states. This periodic oscillation between stable states is an excitation. This state of excitement might be a chemical change, a phase or state change, etc. There is some attribute of the medium itself that changes when it's in an excited state.[23]

We can contrast an excitation with more familiar oscillations in inert media. A common example is a wave in the ocean. A wave is an oscillation and water is the medium that houses the oscillation. Water is an inert medium, so it doesn't have the capacity to become excited. In other words, an oscillation propagating through the water doesn't change anything about the water itself. Water that is part of a wave or not part of a wave is identical. The presence of a wave doesn't do anything to the composition or the state of the water molecules.

Any oscillation is a dynamic disturbance that propagates using some medium. The nature of this dynamic disturbance is a localized region that has more energy than the surrounding medium. As the system evolves towards equilibrium, the local, energetic media distributes the energy to surrounding media. For a wave in the ocean, this is kinetic energy that cycles through the water. Water is pushed into other water, which pushes into other water, etc., and the wave moves across the ocean. The dynamic disturbance is a displacement of water.

In an excitable medium there is something about the media that changes when it's in a state of oscillation. A good example is a non-linear chemical reaction. In contrast to a typical chemical reaction (linear reaction) a solution begins in a state where the two chemicals are separate. When they are combined and make a single solution, the solution evolves *linearly* until the reaction has fully catalyzed. In this reaction there is the beginning, low entropy state, the linear transition, and the final, high entropy state.[13]

A *non-linear* reaction in excitable media might have more than one semi-stable state that can be temporarily sustained as the system evolves. In this type of reaction there might be the initial, low entropy state, the reaction will partially catalyze the solution until it reaches the first semi-stable, but not equilibrium state. We'll call this State A. This is a state of higher entropy, but the reaction hasn't fully catalyzed the solution. In this state there is something about this medium that requires the system to temporarily evolve from this semi-stable, but non-equilibrium state to a state of lower entropy. Obviously, the system cannot continue evolving towards a low entropy state, and eventually the system will reach a different semi-stable state, State B. Again, this new state will be a semi-stable, but non-equilibrium state. The system will repeat the process, temporarily evolving to lower entropy, before continuing to catalyze, and returning to State A. This process will continue, and the system will oscillate between State A and State B until the reaction has fully catalyzed the solution and the system is in equilibrium. Just like in a typical chemical reaction, the solution evolves from low entropy to high entropy, but the evolution of a system of excitable media is non-linear. The system still evolves until it reaches a stable, high-entropy state of equilibrium. The difference is that the system oscillates between more than one semi-stable state as it trends towards higher entropy.[14]

This type of oscillation is very strange. There isn't anything that is part of most people's common experience to serve as a useful analogy to better understand this type of non-linear oscillation. Excitations are inherently dynamical, and there is a limit on how well you can understand something dynamic just by reading a description of it. The best way to gain an understanding of what it means for a medium to oscillate is to watch it happen. There is an excitable, chemical solution that was developed for the purpose of studying its oscillating reaction. This is called the Belousov-Zhabotinsky, or BZ reaction. It is a very profound reaction to watch. It looks like a magic trick, dramatically changing color back and forth every few seconds.

As I'm writing this in the year 2025, I can assume with some safety that you, the reader, might be reading this article on a device that is connected to the internet, or at least have easy enough access to such a device. If that's the case, my suggestion to you would be to watch the video at the web address below.[12] I could go on for pages trying to describe the BZ reaction, but watching the video, seeing this beaker with a solution that is dramatically changing back and forth from a transparent, amber color to a deep, dark blue, oscillating back and forth like magic, seeing it happen in real time will better serve to give the reader an understanding of what an excitable medium is.[4]

`https://www.youtube.com/shorts/ieh9qIkkMJQ`

Thus far, the discussion of excitable media (as well as the video at the address above) has been about oscillating excitable media in general. Forthcoming pages will be focused on specific types of excitations. To use the video as an example, in the video the entire volume of the chemical solution is what oscillates. The whole of the medium oscillates between semi-stable states. Moving forward, we will focus on localized excitations within the volume of the medium.

2.4 Localized Excitations

We have established that particles are excitations in the quantum fields. Stripping away mathematical abstraction we can say that particles are highly localized regions of excitation in the universal vacuum energy. To gain the useful, conceptual understanding of fermions that we are after we need to consider the structures of these excitations. We will quantify the structures in the languages of geometry and topology. We can then distill the important characteristics that become apparent through the math.

Let's return to the oscillating beaker of excitable media that was seen in the video. Even if the solution appears to be changing all at once, it's a chemical reaction that must start at some specific location and spread through the volume. As soon as one molecule undergoes this change it affects the molecules in its immediate vicinity, and the chain reaction causes the solution to appear as though it's changing all at once.

If we imagine taking a slice out of the beaker, it might look like a spot spreading out in a circle until it reaches the edges. Watching that 2D slice long enough for the oscillating medium to complete one cycle, oscillating between one semi-stable state and another, would look something like the cycle in Fig. 6.

In practice the way that scientists look at a slice of excitable media is by putting a thin layer of the BZ solution in a petri dish. The confinement of the oscillation to two dimensions allows the system to develop localized excitations. The result is that at any one moment, different regions of the medium will be in different phases of oscillation.[23]

The reaction happens quickly, but that doesn't mean the medium completely catalyzes from one semi-stable state to the other instantaneously. There is a certain molecular ar-

---

[4]The video cited in the text is owned and uploaded by its cited authors, so it might not be available at any future time. Alternatively, a search on YouTube for 'BZ reaction' will produce many examples that will serve the same purpose.

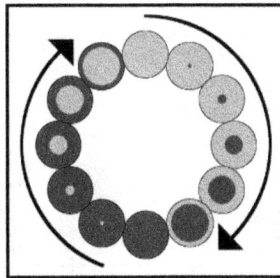

Figure 6: 2D slice of an oscillating chemical reaction shows how oscillations in excitable media are cylces in the state of the media itself

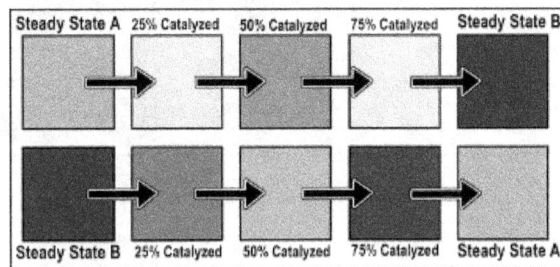

Figure 7: Colors indicate different phases of excitable media catalyzing from one semi-stable state to the other

rangement that is associated with the media being in State A, and another with State B. If the media is in State A and is the reaction is catalyzing to State B, the arrangement of molecular bonds aren't replaced all at once. As the media oscillates from State A to State B there will be different phases, where the molecular structure is to a greater or lesser degree catalyzed to State B. (Fig 7) There will be some location where the molecular bonds in the media at a region are 75% in State A and 25% in State B, etc. (Fig. 8)

There will be some location in the excitable media that the oscillation originates. This is a phase singularity where all phases of the oscillation propagate from. This phase singularity contains all phases of oscillation and is therefore in no particular phase. It is a 0-dimensional point where all phases of the oscillation are simultaneously represented.[6]

In 2 dimensions this phase singularity is called the core. If the core is stationary the resulting wave front will propagate evenly in all directions, represented as rings of differing phases equidistant from the core. Figure 8 shows this happening at a rate where the entire volume of media will fully catalyze from one semi-stable state to the other before the non-linear reaction cycles back to the original semi-stable state. If the excitation cycles at a faster rate, so that as the wave front from one semi-stable state is reaching the outer edge of the petri dish, the core is cycling back to that same semi-stable state, then all phases of the oscillation will simultaneously be represented in the medium. (Fig 9) There will be some parts of the petri dish where all phases from State A to State B are present, and other parts where all phases from State B back to State A are present. When considering the entire petri dish, all phases of the oscillation will exist simultaneously in the medium.

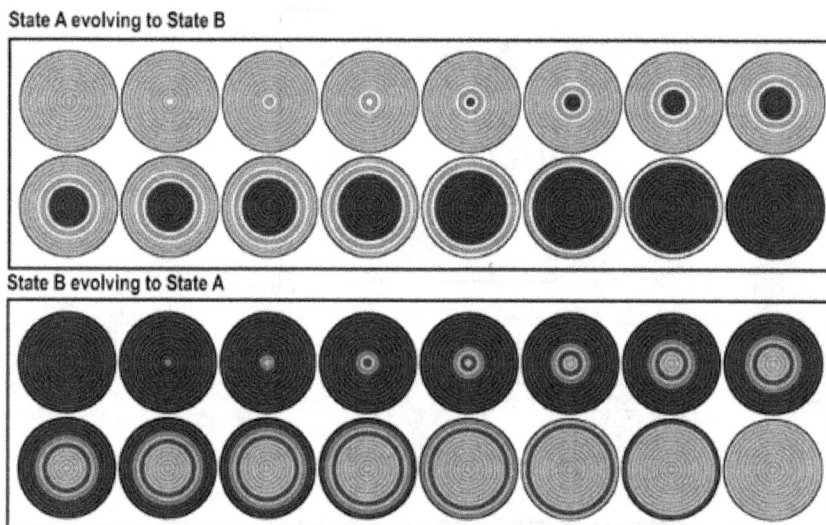

State A evolving to State B

State B evolving to State A

Figure 8: The excitable medium cycles betwen two differnt semi-stable states. The evolution from semi-stable State B to semi-stable State A is not a reversal of the evolution from A to B.

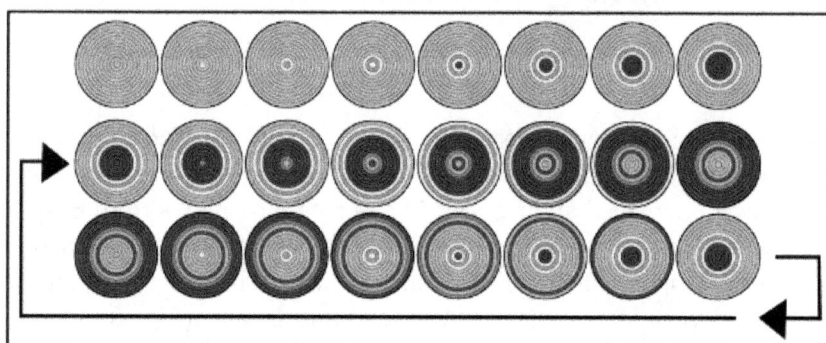

Figure 9: If the oscillating reaction catalyzes between semi-stable states, so that the phase singularity cycles back to a phase of oscillation before the wavefront of equivalent phase reaches the outer boundary, then the petri dish will simultaneously contain all phases of the oscillation. In such a case the medium will oscillate through different configurations of the phases contained in the petri dish.

2D Spiral Waves. If the core is given angular momentum the wave front will propagate in only one direction at a time. The direction of wave front propagation will be constantly changing, rotating around the core. The kind of excitation that propagates from a rotating core in 2 dimensions is called a spiral wave. (Fig. 10) Notice that after the spiral wave has propagated through the media so that every phase of the oscillation is represented in the media, the spiral will then appear to rotate in the petri dish, like a pinwheel.[23]

Important Note on Analogy. The BZ reaction is *continuous*, whereas quantum excitations have *discrete* energy levels. DSSR uses the BZ-type reaction only as *an analogy* of an *excitable system* that can form localized wavefronts. A complete quantum account would include boundary conditions or other mechanisms to ensure discrete particle energy levels.

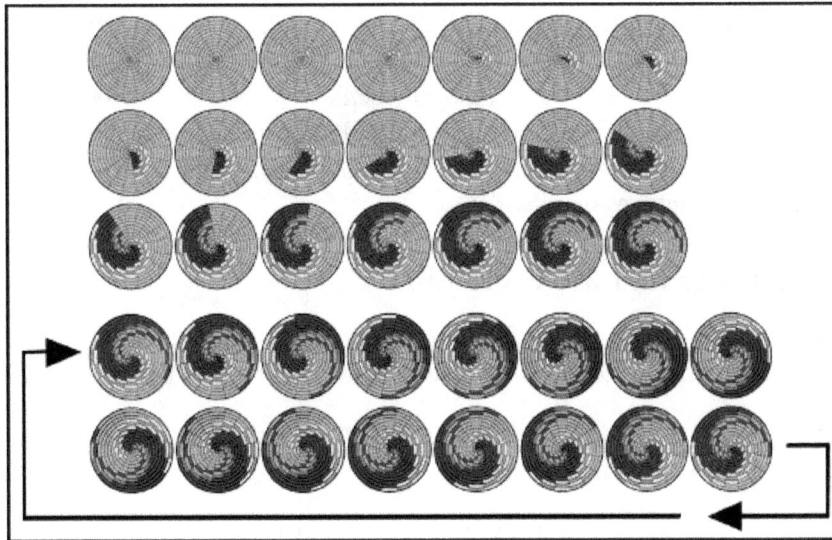

Figure 10: If the phase singularity has angular momentum the resulting excitation will be a spiral wave.

3   Knotted Scroll Waves in 3D

3.1   From Spiral Cores to Filaments

In 2D, excitable media can form spiral waves emanating from a 0D "core." Extending this to 3D yields *scroll waves* emanating from a 1D "filament." These filaments can bend and twist chaotically. Occasionally, they link end-to-end, forming a closed ring.

Scroll Waves.   Scroll waves are the 3-dimensional extension of 2-dimensional spiral waves. As where the core of a 2-dimensional spiral wave is a phase singularity contained in a 0-dimensional point, a 3-dimensional scroll wave propagates from a 1-dimensional phase singularity called the *filament*. A scroll wave is so named because the wave front sort of 'unrolls' as it propagates from the filament, similar to a paper scroll unrolling from a wooden dowel. The simplest kind of scroll wave to imagine is a straight scroll wave, a straight filament with a wave front unrolling out and away from the filament. This kind of scroll wave isn't an a priori impossibility, but a system with conditions that would sustain a scroll wave like this would be very rare. In practice any 3-dimensional system containing a scroll wave is likely to be very dynamic. The forces pushing and pulling on a fragile phase singularity in such a system would cause the filament to writhe and twist and curl, and very often times break apart and/or die off.[23]

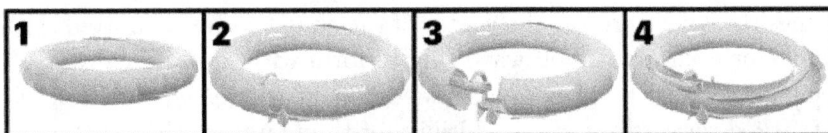

Figure 11: If a filament curls around so that its ends connect to one another the resulting excitation is a scroll wave ring.

225

Scroll Wave Rings. As a filament is curling and writhing, the ends of the filament might curl around and connect to one another. When this happens the resulting oscillator is known as a scroll wave ring. (Fig. 11) Again, this is an oscillation that isn't often built to last. The excitation propagates from different parts of this filament will inevitably send wave fronts towards one another. These wave fronts will crash into each other. The filament will break apart and form new structures or die off.

3.2 Knotted Rings and 3D Constraint

A ring-like filament may *knot* itself, forming stable or metastable structures that cannot easily be untied. Crucially, *topological knots* arise naturally only in **three** spatial dimensions:

- In 2D, you cannot have a nontrivial knot, because lines cannot pass "above or below" each other in a strictly flat plane.

- In > 3 dimensions, knots generally can be untangled or "slipped apart."

- Thus, $3D$ is uniquely suited to stable knotted filaments.

DSSR posits that these *knotted scroll waves* correspond to **fermions**. Because matter is composed of fermions, all matter is thus forced onto a 3-dimensional manifold.

Figure 12: Left and right handed trefoil knots. With the exception of an unknotted ring, trefoil knots are the simplest topological knots.

Topological Knots. In topology, a knot is any closed ring that is threaded through itself in such a manner that it cannot be unwound without being cut.[5](Fig. 12)

Topological knots are uniquely stable in 3-dimensional space. Loops can generally become unwound from one another in >3D. There are methods of "suspending" 3D knots in higher-dimensional spaces. Extending knots to be seen as slices of "knotted" 2-spheres in 4D space, and specifically the boundary conditions of homeomorphic translation to define the entropy exchange mechanism that will be introduced in the next section, is suggested as an area of future research (see §7).

Knotted Scroll Wave Rings. When a scroll wave filament connects to itself so that the filament becomes threaded through the center of the ring it forms a knotted scroll wave ring. It is not possible for a knotted scroll wave ring to form with uniform chirality of wave front propagation.[5] There must be places along the filament where the direction of propagation reverses itself, called perversions of the chirality of the filament. (Fig. 13)

---

[5]A ring, or unknot, is technically a topological knot that doesn't thread through itself.

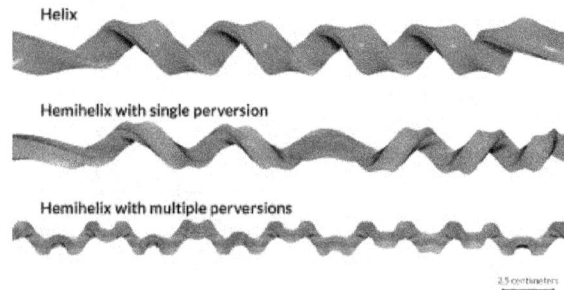

Figure 13: Looking to hemihelices might provide insight into how chirality perversions are formed in knotted filaments as they establish a reference frame, spacetime, that preserves CPT symmetry.[17]

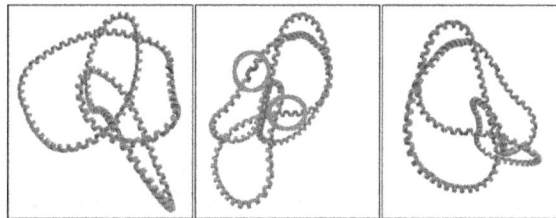

Figure 14: Example of a knotted scroll wave filament with chirality perversions.

It is improbable in a dynamic, turbulent system of an excitable medium that a single filament would form in a way to satisfy the boundary conditions of a knotted scroll wave ring. This kind of oscillation occurs when more than one oscillator, say a pair of unknotted scroll wave rings of opposite chirality, collide with one another. The broken filaments of these oscillators curl and writhe, and the chaotic filaments find stability in coupling with the loose ends of the other oscillator. The pair of oscillators are coupled together, forming a new structure.[23] Whether this is a single excitation or a pair of coupled oscillators becomes an ambiguous matter of perspective.(Fig. 14)

3.3   Why Our 3D? Multi-Slice Possibility

One could ask whether multiple disconnected 3D branes might exist in a higher-dimensional bulk. Even if each local knot is a separate 3D patch, what enforces them to align into one continuous 3D geometry? In typical BZ media, multiple filaments can be misaligned. It might be that in a cosmic setting, boundary or initial conditions could unify local excitations into one large 3D domain. Another possibility is that there is no boundary between 3+1D spacetime and the nD bulk. Our apparent reality *could be* a higher-dimensional space. If matter (fermions) are inherently 3D, and measurements of radiation energy (bosons) obey an inverse square law,[6] there is a possibility that this combination would give the *illusion* of a 3-brane from the perspective of an embedded observer. This remains an open question

---

[6]We suggest the possibility that any measurement of radiation energy performed by embedded observers will record the highest energy density at that spatiotemporal coordinate. In constructing a frame of reference that doesn't violate locality, all embedded observers would measure radiation to obey an inverse square law in any space with 3 *or more* degrees of freedom. This claim will be explored in forthcoming work.

(see §6).

Recap of Sections 2 and 3

Given these premises:

1. All matter in the universe is fundamentally composed of fermions.

2. The existence of a fermion is synonymous with an oscillation that propagates from a phase singularity.

3. Any phase singularity that generates a fermion must have the structure of a topological knot.

4. Topological knots are stable in exactly 3 dimensions.

We are led to the following conclusion:

Matter is constrained to exactly 3 spatial dimensions. The quantum fields might exist in any number of spatial dimensions. Any fermion that exists is being generated by a knotted scroll wave. Fermions will create an embedded 3-brane in any n-dimensional bulk.

4   Entropy, Local vs. Global

A core DSSR claim is that a fermion-generating filament remains in a *low-probability (low-entropy)* state, but does *not* violate the second law of thermodynamics. We must clarify how.

4.1   Fermions as Localized Regions of Low Entropy

Fermions are fundamental particles, but they are also excitations of the vacuum energy. Fermions are localized regions of the bulk where an oscillation has become wrapped on itself so that a quantum value of energy is bound up in the form of matter. In short, a fermion is an oscillation.

Just like any other oscillation, we are able to consider a fermion as a system in a state of oscillation. We do not question the fundamentality of fermions. What we are asserting is that fermions are fundamental particles that exist in the framework of spacetime. It seems likely that there exists a bulk that extends beyond what is observable to us, as observers made of fermions, observers embedded in spacetime.[21] Fermions are localized regions of the bulk where undulating, rippling currents of the vacuum energy have created a 1-dimensional phase singularity that has knotted itself and become embedded in 3 dimensions. Without having definitive knowledge of the dimensionality of the bulk we can be certain that the knotted phase singularity can be fully described in a coordinate system with 3 axises. The fluctuating vacuum energy in this coordinate system could be described by considering the region as a phase space, an exitability gradient, or any other suitable method of quantifying the local behaviors of the vacuum energy as they pertain to the knotted filament. Our point is that just like any arbitrary region of space could be defined as a physical system, we should be able to define any arbitrary region of the bulk and do the same.

In considering the region of the bulk that houses a fermion excitation we can compare the unique configuration of that 'system' to other possible configurations:

- Given the distribution of matter and energy in the universe, it is less likely the a region of the bulk would contain the energy spike required to generate a fermion. Imagine we were to divide the entire universe into 'quark sized' cubes and pick one of those cubes at random. There is a certain, quantized threshold of energy needed for there to be an excitation that generates a fermion. Considering only the amount of energy contained in those cubes, there would be a much greater probability that the cube selected at random will not contain enough energy for the kind of excitation that generates a fermion.

- If we grant that there is an excitation, we can consider the likelihood of that excitation having the structure required to generate a fermion.

  - If an excitable medium is in a state of oscillation, it is less likely that the phase singularity the excitation propagates from curls around on itself so that the ends of the filament connect with one another. There are more possible configurations of a straight scroll wave than there are possible configurations of a scroll wave ring.
  - If the ends of the filament were to connect with one another and form a closed ring, it is less likely that the filament would thread through the center of itself, knotting itself with necessary chirality perversions. There are more possible configurations of a closed ring that is unknotted than possible configurations of a knotted filament.[7]

The existence of any fermion in the universe is synonymous with the existence of an oscillation propagating from a knotted filament with chirality perversions. These are the boundary conditions of the oscillating system that generates a fermion. Given all possible configurations of the system, the oscillating system that satisfies these boundary conditions is in a state of low probability.

It's not just that the existence of a fermion corresponds to an oscillation in a low probability state, but that the fermion is synonymous with a dynamical, oscillating system that evolves from one unique configuration to another. The existence of a fermion requires that each and every unique configuration the oscillating system evolves to must satisfy the boundary conditions to generate a fermion. Every configuration of a system that generates a fermion is a system in a state that has a low probability of occurrence.

States of low probability are states of low entropy. As the system evolves over time, it evolves exclusively to states of low entropy.

---

[7]Truly, there are an infinite number of configurations of a both knotted and unknotted rings. Given similar boundary conditions, ie spatial constraints, total length of the ring, etc, there are more possible configurations of an unknotted ring than a knotted ring with chirality perversions.

4.2 Subsystem vs. Universe

Thermodynamics forbids *global* entropy decrease in a closed system. It does *not* forbid local decreases if they are offset by an even greater entropy increase in the environment.

- **Ice Formation Analogy:** Water freezing forms a low-entropy solid, but the surroundings gain heat. The net entropy still rises.

- **Knotted Filament:** The local region containing a stable knot has fewer microstates available than an unstructured swirl. This begs questions regarding what entropy might be from a perspective outside of spacetime, or a perspective that accounts for the dynamics of systems in the nD bulk. It may be the case that global vacuum can dissipate energy or produce more random excitations elsewhere, so total entropy goes up. Entropy establishes the arrow of time for observers embedded in spacetime. 4D spacetime entropy might not have a 1:1 correspondence when generalized to a reference frame with more degrees of freedom.

Does Spacetime Embedded Entropy Have a 1:1 Correspondence to Systems in the nD Bulk? DSSR does not assert any faults in the current description of entropy. What is being asserted is that there is an equivalent way to look at entropy that gives an alternative description of the way the arrow of time is established. The current model considers fermions to be truly fundamental point particles with no inherent entropy value. When the particles interact with one another in a system, they will become more disordered over time, and the entropy of the system will increase. The model DSSR proposes says that the 3D framework is established by topological constraints on excitations that generate fermions. Having established the boundary conditions of an oscillation that generates a fermion, we can adopt a parallel perspective where fermions are endowed with some inherent value of entropy. We are not advocating a change in the way entropy is quantified. We are proposing the inclusion of additional information, the additional term accounting for the low entropy states of the oscillations that generate fermions, when considering the total entropy of material systems embedded in spacetime.

It is this caveat, that the systems are embedded in spacetime, that leaves some ambiguity regarding the 'dual perspective' approach to quantifying entropy. The ambiguity is introduced as follows:

1. The oscillations responsible for fermion generation are endowed with a structure that necessitates fermions being confined to 3 spatial dimensions.

   (a) We will take for granted that this plays a causal role in the 3D nature of the macro world.

2. This structure is the result of the dynamics of a higher-dimensional excitable vacuum energy.

3. Fermions are the particles that experience time.

4. Fermions are the massive particles that distort the geometry of spacetime. (More on this in §5.)

5. Considering (1-4), we take for granted that without fermions there isn't any reference frame that can sensibly called "spacetime."

6. Entropy is a measurement of the number of equivalent microstates of a physical system, based on the current state of the system.

   - The idea of measuring entropy is based on considering the state of a system at a single moment in time. It is the idea of taking a static, momentary 'snapshot' of a physical system, frozen in 3 spatial dimensions, and comparing that state to other possible arrangements of the fermions that make up that system.

   - In order to measure the entropy of a system, there must be a material system, made of fermions, to be measured.

7. It is generally accepted that increasing entropy plays a causal role in establishing the arrow of time.

8. Entropy is a value that increases over time.

With (1-8) in mind, we can make the assertion that entropy has a dependent relationship on the existence of fermions. Entropy is the measurement of the configuration of a physical system. Without fermions there are no physical systems. Spacetime also has a dependent relationship with existence of fermions. Without fermions there is no frame of reference that can rightly be called spacetime.

Quantifying the entropy of an oscillating system that generates a fermions is not as straight forward as quantifying the entropy of macro systems. If there is a method of accounting for the entropy of oscillations responsible for fermion generation, it would likely require that entropy is considered as an interdependent relationship with those oscillations. In contrast to a dependent relationship, where the existence of fermions is prerequisite to entropy, an interdependent relationship is one where the relationship between things has causal power in endowing each relata with the properties that make them what they are.[8]

For the purposes of this paper, we will circumnavigate the issue of what significance we can place on spacetime embedded entropy in systems that exist in an n-dimensional bulk by qualifying that *if systems structurally equivalent to knotted, fermion-generating excitations were to be recreated macroscopically, embedded in spacetime*, that these systems would be seen as evolving to states of lower entropy.[9]

---

[8] An example of an interdependent relationship is that of a teacher and a student. Without a student to teach, one cannot be a teacher. Without a teacher to study, one cannot be a student.[11] Formalizing this interdependent relationship will be suggested as an area of future research in §7.

[9] If these systems were to be recreated as macroscopic systems embedded in spacetime any apparent localized entropy decrease would of course be off set by increasing entropy in the environment that contains the localized system. Our point is not to challenge the 2nd law. We only suggest a parallel view of entropy that might include fermion-generating excitations of vacuum energy.

Section 4 Recap

Given these premises:

1. The oscillation that generates a fermion can be viewed as a physical system with a quantifiable value of entropy.

2. For a fermion to exist the oscillation generating the particle must propagate from a knotted filament with chirality perversions.

3. The continuous existence of a fermion dictates that the system generating the fermion evolves exclusively to states that satisfy these boundary conditions.

4. Considering all possible configurations the system might be in, any configuration where a knotted, twisted filament exists for a scroll wave to propagate from should be considered a configuration with a low-probability of occurrence.

We should conclude:

From this perspective, the evolution of a system that generates a fermion from one state where the particle exists to another state where the particle continues to exist can be seen as the system evolving to states of lower entropy.

5 Emergence of Time

5.1 Mass, Geodesics, and Time

In relativity, *massless* bosons travel on null geodesics at c and experience no proper time (they cannot be brought to rest). Conversely, *massive* fermions follow timelike geodesics, experiencing the familiar flow of time. DSSR extends this to highlight that *localized* excitations (e.g. knotted filaments) define reference frames where time passes. Particles have internal states that evolve—they exchange energy, interact with the environment, and do so along a thermodynamic arrow of time.

Gravity and the Emergent Temporal Dimension. Gravity is the force that distorts spacetime. Gravity doesn't pull massive bodies towards one another, it distorts the geometry of spacetime in the presence of massive bodies. Bodies accelerate towards their collective center of mass because the geometry of spacetime dictates this to be the path of least resistance.

Mass is the property that causes a body to resist acceleration through space and subjects a body to gravitational acceleration. Another way of saying this is that mass is the property of a particle that enables it to distort the geometry of spacetime [9].

Massless bosons do not resist acceleration through space, so they are always moving at the speed of light. Because they are moving at the speed limit of the universe they aren't subjected to gravitational acceleration. In contrast, fermions are endowed with mass and resist acceleration through spatial dimensions, while being subjected to gravitational acceleration in the temporal dimension. Only particles with intrinsic mass experience time.

Fermions are the particles that are endowed with mass, and therefore the particles that experience time. Fermions are also the particles that all matter is made of.[25] Any physical system that exists is made of matter.

5.2 Entropy-Driven Arrow

The second law states that *global* entropy tends to increase. As any macroscopic system evolves, it drifts toward higher-entropy states. Fermions are building blocks of that system, but each fermion's local configuration remains relatively low-entropy. The *net* effect is consistent with the typical arrow of time: systems made of many fermions reach equilibrium or more disordered states overall, while individual fermions remain topologically stable. This *"local low-entropy, global high-entropy"* interplay naturally creates a unidirectional sense of time for embedded observers.

The Probabilistic 2nd Law. It's generally accepted that the arrow of time is established by the 2nd law of thermodynamics. As physical systems evolve over time they will trend towards higher entropy. There is no rigid, cause and effect reason that this happens. It is an overwhelming probability. Particles in a system move and bump into each other and exchange energy. As the system evolves from one unique arrangement of its particles to the next there is some degree of randomness in how it evolves. There are more possible ways that the particles of a system can be arranged in a state of equilibrium than there are possible ways that a system can be arranged in any other specified state. Therefore, as the system evolves there is an overwhelming likelihood that it will evolve towards equilibrium.

What's more, as a system moves closer to equilibrium, it becomes increasingly more likely that the next unique arrangement of the system will be closer to equilibrium than the previous state. Equilibrium states are highly disordered. The closer to equilibrium a system is, it becomes increasingly less likely it is that the system would move to a more ordered state.[19]

Section 5 Recap

Given these premises:

1. For any physical system to exist the fundamental components of that system must also exist.

2. Any physical system that can be observed to exist in the universe must include fermions as fundamental components of that system.

3. We have established that as fermions exist over time, they can be viewed as fundamental systems evolving to states of decreasing entropy.

4. As any macroscopic system evolves to states of higher entropy, obeying the 2nd law of thermodynamics and in accordance with the arrow of time, the fundamental components of that system are evolving to states of lower entropy.

We should conclude:

The arrow of time can be seen as the exchange of entropy from the particles that a system is made of, to the system that is made of those particles.

Rather than holding the view that entropy increases in any closed system, we can adopt the view that as any physical system evolves macroscopically to a state of increased entropy, the fundamental components of the system evolve to states of low entropy. When accounting for the decreased entropy of the fundamental components of any macroscopic, physical system, we should conclude that entropy can be seen equivalently as a value that is conserved over time.

6 Open Questions and Future Work

Why Our 3D? If 3D is special for stable knots, is it possible that multiple 3D slices exist in an *n*-dimensional bulk, each forming separate low-dimensional pockets of matter? If so, why do we observe *one* unified 3D continuum rather than many disjoint 3D domains? DSSR suggests there might be boundary conditions or global constraints that unify local singularities. A thorough demonstration remains a topic for future study.

Alignment of Multiple Filaments. Even if each fermion emerges from a local topological singularity, the question remains how these local 3D excitations *all align* to form our shared 3D geometry. In chemical systems like BZ, multiple scroll-wave filaments can remain separate or misaligned. We hypothesize that cosmic boundary/initial conditions might *couple* the excitations into one large 3D manifold. Providing a more rigorous mechanism is suggested as a direction of future research.

Discrete vs. Continuous. The BZ reaction is continuous, whereas quantum energy levels are discrete. DSSR uses BZ-type phenomena as an analogy for *excitability*, not to claim a perfect 1:1 match with quantum behavior. Detailed PDE or topological quantum field theories might demonstrate how discrete fermionic excitations arise from a continuum vacuum under the right boundary conditions.

Gauge Interactions. We have not explored how gauge fields $(SU(3) \times SU(2) \times U(1))$ fit into the knotted filament picture. One possibility is that *topological quantum field theories* (TQFTs) might embed both the knot-based matter sector and gauge interactions in a single mathematical framework.

7 Conclusion and Outlook

7.1 Summary of Main Claims

1. **3D from Knots.(§2 and §3)** Knotted filaments are stable only in exactly three spatial dimensions, explaining why matter (fermions) localize in 3D if they arise from such topological excitations.

2. **Excitable Vacuum.** The quantum vacuum is recast as a dynamic, excitable background, analogous to but not exactly like chemical excitable media.

3. **Local Low Entropy.(§4)** The excitation that generates a fermion is a knotted, twisted scroll wave ring. The boundary conditions of this type of excitation require the local environment to be in a low-entropy configuration. Since all configurations that the local environment could be in must have a knotted scroll wave, the system will evolve locally to only those low-entropy configurations. Each fermion is a local, low-entropy arrangement; globally, entropy still increases, preserving the second law.

4. **Entropy exchange mechanism.(§5)** Excitations that generate matter particles are in low entropy configurations, but macroscopic, material systems evolve to states of high entropy. This relationship can be described as an *entropy exchange mechanism*. Adopting this perspective, the entropy exchange mechanism plays a causal role in establishing the arrow of time.

5. **Emergent Spacetime Model.** From a high-dimensional space that houses the dynamic, undulating sea of vacuum energy, spacetime emerges in 2 distinct steps: (1) Fermions create a 3-brane, establishing the **space** of spacetime. (2) The boundary conditions of a knotted scroll wave require that fermion-generating excitations evolve to low-entropy configurations. As material systems evolve towards equilibrium, the arrow of **time** emerges from the entropy exchange mechanism. **Spacetime** emerges from n-dimensional vacuum energy.

7.2 Future Steps

Below are suggested directions to further develop DSSR:

- *Formal TQFT Approach.* One can attempt a rigorous topological quantum field theory in which knot-like solutions are spin-1/2 excitations, bridging standard QFT and the geometry of knots.

- *Experimental/Computational Models.* Realistic PDE simulations of 3D scroll waves or potential lab analogs might yield partial support for topological constraints in excitable media.

- *Discreteness from Continuum.* A deeper quantum boundary mechanism might be needed to reconcile discrete particle spectra with the continuum nature of classical excitable waves.

- *Formalizing Interdependence via Autoevolute Curves.* In differential geometry, the evolute of a curve is defined as the set of all points that are the center of curvature for all points along the curve. In other words, if you were to follow along a curve and at every infinitesimal point, map another point that is the center of curvature for that location on the curved line, those points would make a continuous line or shape. That shape would be an envelope containing all normals of the original curve.

Figure 15: [20]DSSR's entropy exchange mechanism might be formalized by considering a subset of 3D knots with the boundary condition that they must be self-congruent autoevolute curves, and that as these curves embed themselves in 3D space they generate a reference frame that preserves CPT symmetry.

The curves in Figure 15, named *autoevolute curves*, are their own evolutes. Every point on each curve can be mapped as the center of curvature to another point on the curve.[15]

Formalizing the entropy exchange mechanism proposed in DSSR will require that boundary conditions for possible configurations of fermion-generating vacuum energy excitations be well defined. In turn, formalizing the DSSR's entropy exchange mechanism is prerequisite to formalizing interdependence of spacetime embedded entropy and knotted filaments in a higher-dimensional configuration space. Looking to auto-evolute curves to formalize the duality required of knotted excitations of fermion fields seems to be a logical approach to establishing these important boundary conditions.

While it's unlikely that every fermion-generating knot could be mapped 1:1 to its own evolute in 3D, a likely possibility is that with additional degrees of freedom any non-trivial 3D knot could be mapped to an *n*-dimensional autoevolute curve.

Looking to autoevolute curves that satisfy the condition of being a 3D knot to establish a CPT symmetric reference frame is a promising area of research to establish these necessary boundary conditions. This is a vital 'next step' for DSSR.

- *Gravitationally Mediated Entanglement in the CPT Symmetric Cosmological Model.* Methods that have led to quantization of other fundamental forces, if they were to be applied to the gravitational force, would need measurements from a very sensitive measurement apparatus, capable of detecting the much weaker force carrier of the gravitational force. Any measurement apparatus would need to be sensitive to the changes in the momentum of massive bodies as a graviton is transmitted between them. Trying to determine exactly how sensitive the detector would need to be shows that the mass and spatial separation of the particles being measured has a value equivalent to the Schwarzchild radius.[8]The gist of it is that when we calculate how sensitive a measurement apparatus would need to be to detect a single graviton, what we discover is that any measurement would find itself hidden behind an event horizon.

  Any apparatus measuring changes in the *momentum* of particles, by measuring changes in spatial coordinates over a period of time, would be hidden from a spacetime embedded observer. Recent work (which was also presented at the Minkowski Institute's 7th Spacetime Conference) proposes the possibility that spacetime embedded observers might be able to measure fluctuations in the *temporal* dimension by precise measurement of *Proper Time Oscillators*.[27]

236

Observation of gravitational entanglement would serve as evidence that gravity must be quantum and not classical, as this interaction would require a quantized force carrier to facilitate entanglement. The dynamical symmetry of knotted filaments is reminiscent of entanglement. If we hypothesize that knotted phase singularities are endowed with their unique structure via *gravitationally mediated entanglement* that is *prerequisite* to establishing a *CPT symmetric* reference frame, we can look to a 4D geometric derivation where we might be able to measure changes in momentum by observing temporal fluctuations.

Assuming real particles act as proper time oscillators, we should consider them in the context of *The CPT Symmetric Model* of the universe. [2] For an observer embedded at the spacetime coordinate of a proper time oscillator, we could construct a geometry where, locally, there would be no distinction between *space like boundaries* and *time like boundaries*. Establishing relativistic limitations of this spatiotemporal locality, for an observer embedded at that spacetime coordinate, for *early time observers* near that coordinate, and for *late time observers* near that coordinate, would approach some limit that could be quantified as a spacetime vector.

Comparing these vectors to possible values of momentum vectors representing graviton transmission could serve as evidence of *gravitationally mediated entanglement*, evidence unavailable by direct measurement to spacetime embedded observers.

We hesitate to say this this is a possible future research direction *for DSSR*. This does appear to be a very interesting avenue of future research, but the only potential contribution that DSSR could offer is the model of fermions as knotted phase singularities that establish a CPT symmetric reference frame, which is consistent with fermions being modeled as gravitationally entangled oscillators. This is a minor contribution. If there are any insights to be gained via this avenue, the development of **Proper Time Oscillators** and **The CPT Symmetric Model** will be what allows for these insights. This is an area of research with profound potential.

- *DSSR and the Hard Problem of Consciousness* The aspects of DSSR that have been presented in this article have been sort of "reverse engineered" from an initial exploration of knotted scroll waves as a potential basis for a monist or panpsychist model of consciousness.

  Developing DSSR as an ontology that can account for the inclusion of consciousness in the set of things that exist in the universe was the original motivation for the work that led to this theory. This remains the ultimate goal in the development of DSSR.

DSSR thus unifies wavefunction-based realism and a *3+1D* emergent spacetime vision through a *topological* and *entropic* lens. We hope this prompts new research—both mathematical and experimental—into the nature of these elusive knotted excitations.

Acknowledgments

The peer reviewer for this paper went above and beyond. I am an outsider to the world of academia, and (until now) I was never in a position where I needed to learn to write in a style typical of original research papers. The anonymous reviewer did a great deal to help me make this paper less jarring to its intended audience. This person really should be listed as a co-author of this work. Whoever you are, I am very grateful for your help.

On that note, the editor of this volume, Kyley Ewing, has been very helpful to me in navigating the unfamiliar world of academia.

Thank you to Joey, Sully, Chris, Shane, friends and family that have let me bounce ideas off you.

Thank you to friends and family who made financial contributions which allowed me to travel to Bulgaria to present these ideas at the Minkowski Spacetime Conference. Thank you to the Cass Clay chapter of the Awesome Foundation who awarded me a grant for the same.

Thank you to The Minkowski Institute for being open to perspectives originating from all avenues, and for fostering a collaborative, encouraging culture that holds paramount the goal of advancing our understanding of the natural world.

Brynna, Ruby, Otis, Waylon, thank you to my wonderful family for your love and support.

References

[1] Ainsworth, P., *What is Ontic Structural Realism?*, *Studies in History and Philosophy of Science Part B 41(1), 2010.*

[2] Boyle, L., Finn, K., Turok, N., *CPT-Symmetric Universe*, *Physical Review Letters 121(25), 2018.*

[3] Coleman, S., *Fate of the False Vacuum: Semiclassical Theory*, *Physical Review D 15(10), 1977.*

[4] Cush, *The Standard Model of Elementary Particles*, *Wikipeida, 2017.* `https://en.wikipedia.org/w/index.phptitle=File:Standard_Model_of_Elementary_Particles.svg&dir=prev`

[5] Dähmlow, P., Alonso, S., Bär, M., Hauser, M., *Twists of Opposite Handedness on a Scroll Wave*, *Physical Review Letters 110(23), 2013.*

[6] Deepa, Sankar. *Fascinating Phase Singularities Physics JCE. YouTube, 2023.* `https://www.youtube.com/watch?v=LqdKH8xMMgQ`

[7] Dirac, P., *The Principles of Quantum Mechanics*, *Clarendon Press, 1930.*

[8] Dyson, F., *Is a Graviton Detectable?*, *International Journal of Modern Physics A 28(25) 2013.*

[9] Einstein, A., *On the General Theory of Relativity*, *Sitzungsberichte Preuss. Akad. Wiss. Berlin (Math. Phys.), 1915.*

[10] Feynman, R., Leighton, R., Sands, M., *The Feynman Lectures on Physics, Pearson, 2013.*

[11] Hong, C., *Personal Communication, Spet, 2024.*

[12] K, B. *Classic Belousov-Zhabotinsky Reaction, Youtube, 2024.* `https://www.youtube.com/shorts/ieh9qIkkMJQ`

[13] Kaplan, D., Glass, L., *Understanding Nonlinear Dynamics, Springer, 1995.*

[14] Kapral, R., Showalter, K. (Eds.), *Chemical Waves and Patterns, Kluwer, 1995.*

[15] Karcher, H., Tjaden, E., *ccc-Autoevolutes arXiv preprint arXiv:2102.00832 (2021).* `https://arxiv.org/abs/2102.00832`

[16] Ladyman, J., *Structural Realism, Stanford Encyclopedia of Philosophy (Summer 2023 Edition).*

[17] Liu, J., Huang, J., Su, T., Bertoldi K., Clarke, D.R. *Structural Transition from Helices to Hemihelices. PLoS ONE 9(4) e93183 (2014).* doi:10.1371/pone/journal.pone.0093183.

[18] Ney, A., *Three Arguments for Wave Function Realism, European Journal for Philosophy of Science 13(50), 2023.*

[19] O'Dowd, M., *PBS Space Time Series on Vacuum Energy, Quantum Mechanics, etc., Public Broadcasting Service (2017–2018).*

[20] Palias, R.S., Karcher, H., *3DXM Virtual Math Museum, 3DXM Consortum (2004-2024.* `https://virtualmathmuseum.org/index.html`

[21] Randall, L., *Warped Passages: Unraveling the Mysteries of the Universe's Hidden Dimensions, Harper Perennial, 2006.*

[22] Schwartz, M., *Quantum Field Theory and the Standard Model, Cambridge University Press, 2013.*

[23] Strogatz, S., *Sync: The Emerging Science of Spontaneous Order, Penguin, 2004.*

[24] Susskind, L., *The Black Hole War: My Battle with Stephen Hawking to Make the World Safe for Quantum Mechanics, Little, Brown, 2009.*

[25] Weinberg, S., *The Quantum Theory of Fields, Cambridge University Press, 1996.*

[26] Weingard, D., Steinbock, O., Bertram, R., *Expansion of Scroll Wave Filaments Induced by Chiral Mismatch, Chaos 28(4), 2018.*

[27] Yau, H., *Quantum Harmonic Oscillator in Time, 2025.* ffhal-04924836ff `https://hal.science/hal-04924836v1/document`

www.ingramcontent.com/pod-product-compliance
Lightning Source LLC
Chambersburg PA
CBHW051208200326
41519CB00025B/7050